CLIMATE: SERIOUSLY?

A Half-Hopeful, Half-Terrified Guide to What's Really
Going On (And What We Can Do About It)

By Scott Danville

**DANVILLE
HOUSE**
BOOKS
Words That Matter

Published by Danville House Books,
an imprint of Danville House Publishing

ISBN: 979-8-9986320-2-0

Printed in the United States of America

First Edition

Dedication

To Ray Johnson, whose tireless dedication to climate science education through the Institute of Climate Studies USA has illuminated the path for countless individuals, including myself. Your ability to translate complex climate science into accessible knowledge while maintaining scientific integrity has been the guiding light for this book.

And to Lola Johnson, whose unwavering support enabled Ray's work to flourish. Behind every climate communicator stands someone who keeps the home fires burning (using renewable energy, of course). Your partnership reminds us that addressing climate change requires not just scientific understanding but also human connection and mutual support.

And for all my students, Past, present, and future—you are why this book exists. Your questions challenge me, your passion inspires me, and your future keeps me honest. You taught me that the most important climate metric isn't parts per million of carbon dioxide, but parts per million of hope, determination, and courage. This book is my attempt to bottle some of that spirit and pass it forward.

May you always know that your actions matter in ways both measurable and immeasurable, that complexity is no excuse for complacency, and that the most powerful force for planetary health is people who refuse to give up.

To all who seek to understand our changing planet with both intellectual rigor and human compassion—this book is for you.

The future is unwritten, and you hold the pen.

Table of Contents

Table of Contents

Prologue

Between Hope and Concern: A Teacher's Guide to Climate Change

Climate change is the defining challenge of our time—a test of our scientific understanding, our political systems, our economic models, and our moral imagination. Understanding it requires both rigorous science and accessible explanation. My hope is that this book builds bridges between climate science and public understanding.

Because in the end, climate science isn't just about measurements and models—it's about our shared home and shared future. And that's a story worth telling correctly.

As a teacher facing classrooms full of increasingly concerned students, I also simply wanted to answer their questions honestly without sending them spiraling into despair. "Is it really as bad as the news says?" "Are we already doomed?" "What can someone like me actually do about something so huge?"

These weren't just academic inquiries—they were existential ones, asked by young people inheriting a planet undergoing transformations unprecedented in human history. I watched their expressions shift between fear, anger, confusion, and determination as we discussed rising temperatures, melting ice sheets, and increasingly extreme weather.

What struck me most was not their anxiety (though that was certainly present), but their hunger for straight talk without sugar-coating or fatalism. They didn't want climate bedtime stories that downplayed the crisis, nor did they want apocalyptic narratives that rendered action pointless. They wanted the complicated, messy truth and practical pathways forward—a roadmap for navigating terrain that grows more challenging by the year.

This book grew from those conversations and from my own journey trying to make sense of climate science, policy, psychology, and activism. It's not a definitive text—no single volume could encompass something as complex as climate change—but rather an attempt to provide useful coordinates for anyone trying to understand where we are, how we got here, and what we might do next.

I've structured this book to mirror how I've come to think about climate change: first understanding the problem in all its scientific complexity, then examining its cascading impacts across natural and human systems, then exploring the solutions emerging at every scale from individual to international. Throughout, I've tried to balance sober assessment with genuine grounds for hope, technical accuracy with human accessibility, and urgency with patience for the long road ahead.

Climate change represents perhaps humanity's greatest challenge, but also an unprecedented opportunity to reimagine our relationship with each other and the planet that sustains us. How we respond in the coming decades will echo for centuries, shaping the world that future generations will inhabit. This responsibility is enormous, but so is our capacity for creativity, cooperation, and change when we need it most.

This book won't tell you what to think about climate change, but it aims to help you think more clearly and act more effectively in a world being reshaped by it. My hope is that after reading, you'll feel better equipped to engage with climate challenges in ways that match your unique circumstances, skills, and spheres of influence—not because any one person can "solve" climate change, but because our collective response emerges from millions of individual contributions woven together.

In the pages that follow, we'll explore climate science without unnecessary jargon, examine impacts without apocalyptic framing, and investigate solutions without technological magical thinking. We'll look at the psychological, social, and political dimensions alongside the scientific and technical ones. And throughout, we'll maintain that essential balance between clear-eyed recognition of the challenges and authentic hope based on what humans are already doing to address them.

Welcome to a conversation about climate that's both half-hopeful and half-terrified—because that's where the truth lives, and that's where the work begins.

Chapter 1: The Greenhouse Effect Is Not a Spa Treatment

Why CO$_2$ Is Turning Our Planet Into a Slow Cooker (Minus the Pot Roast)

Have you ever walked across a scorching parking lot in summer, opened your car door, and been greeted by a blast of air so hot it feels like you've opened a portal to the sun's surface? That moment when the seatbelt buckle brands you like a reluctant cow? Congratulations! You've just experienced the greenhouse effect—nature's way of saying, "You should have cracked a window."

This isn't just an unfortunate coincidence between your poor parking choices and the laws of physics. Your car heated up because sunlight easily passes through your windows, gets absorbed by the seats and dashboard (often in heat-loving dark colors because car designers apparently went to the Darth Vader School of Interior Design), and converts to heat that can't escape as easily as it entered. Physics: 1, Your Comfort: 0.

Earth works the same way. Instead of glass windows, we have an atmosphere filled with certain gases that do the same job. They let sunlight in but play bouncer when heat tries to leave by increasing the concentration of greenhouse gases in the atmosphere, we're amplifying the planet's natural greenhouse effect and turning up the dial on global warming. These "greenhouse gases" include carbon dioxide, methane, water vapor, and a few others with names that sound like they should be prescribed for moderate to severe plaque psoriasis.

The Science: Not As Complicated As Your Dating Life

For most of human history, this system worked beautifully—a natural thermostat keeping our planet at a comfortable average temperature of about 15°C (59°F). Without this natural greenhouse effect, Earth would be a frozen wasteland averaging -18°C (0°F), and we'd all be popsicles with student loan debt.

But here's where we humans decided to get creative. Since the Industrial Revolution, we've been pumping extra greenhouse gases into the atmosphere, mainly by burning fossil fuels that had been safely sequestered underground for millions of years. It's like we found Earth's thermostat and collectively decided, "What this perfectly habitable planet needs is a fever!"

The sun bombards Earth with energy in the form of light. About 30% of this energy bounces right back into space, reflected by clouds, ice, and your neighbor's blindingly white cruise ship. The rest gets absorbed by land, oceans, and the atmosphere, warming them up like a cosmic microwave.

As Earth's surface warms, it releases heat in the form of infrared radiation. If we could see infrared light, the ground would look like it's constantly glowing, similar to everyone leaving a hot yoga class.

Here's the crucial part: greenhouse gases in the atmosphere absorb heat radiating from the Earth's surface and re-release it in all directions—including back toward Earth's surface. Without carbon dioxide, Earth's natural greenhouse effect would be too weak to keep the average global surface temperature above freezing. This traps heat energy in the lower atmosphere, warming the planet further. It's like adding extra blankets on a cold night. One blanket is cozy; four blankets have you waking up at 3 AM contemplating the life choices that led to your current sweat-soaked predicament.

Carbon Dioxide: The Houseguest Who Won't Leave

Carbon dioxide (CO_2) gets the most attention in this climate drama, and for good reason. It's not the most powerful greenhouse gas, but it's:

1. Remarkably persistent (staying in the atmosphere for hundreds of years, like that embarrassing social media post from college that potential employers keep finding)
2. Being produced in massive quantities by human activities
3. At its highest concentration in at least 650,000 years, with global average atmospheric carbon dioxide reaching 419.3 parts per million in 2023, a new record high

Before the Industrial Revolution, atmospheric CO_2 levels hovered around 280 parts per million (ppm). Today, we're well over 420 ppm and climbing faster than your heart rate when the barista gets your complicated coffee order exactly right. That's an increase of 50% since pre-industrial times—the atmospheric equivalent of gaining 50% of your body weight since your great-great-grandmother was born. Earth didn't even get a chance to buy new pants.

Methane deserves dishonorable mention too. Though it has a shorter lifetime in the atmosphere than CO_2, methane is around 80 times more powerful at trapping heat in the first 20 years after release. It comes from sources like livestock digestion (yes, cow burps and farts are altering our climate—sometimes the apocalypse comes with a punchline), rice paddies, natural gas leaks, and decomposing landfill waste. According to the Center for Climate and Energy Solutions, methane concentrations have more than doubled since pre-industrial times, adding significantly to our planetary fever.

A particularly troubling aspect comes from thawing permafrost in Arctic regions. As these permanently frozen soils warm up, ancient organic matter begins to decompose, releasing methane and carbon dioxide. It's like finding that container at the back of your freezer from 2017, except instead of just ruining your appetite, it helps ruin the climate.

Earth's Energy Budget: When Accounting Actually Matters

To understand climate change properly, we need to think about Earth's energy balance—what scientists call the "Earth's energy budget." It's essentially accounting, but for heat instead of money, and unlike your personal finances, you can't just ignore it until the next paycheck arrives.

Here's how the ledger breaks down:

Energy coming in: About 340 watts per square meter of solar radiation reaches Earth's upper atmosphere.

- Roughly 100 watts get reflected back to space (by clouds, snow, ice, and light-colored surfaces)
- The remaining 240 watts get absorbed by land, oceans, and the atmosphere

Energy going out: To maintain a stable temperature, Earth must radiate 240 watts per square meter back to space.

- Some escapes directly to space
- Some is absorbed by greenhouse gases and re-emitted in all directions
- The balance determines whether Earth warms or cools

When greenhouse gas concentrations increase, less heat escapes to space, so the system warms until a new balance is reached. It's like putting on a sweater without adjusting the thermostat—your body generates the same heat, but less escapes, so you get warmer.

What makes this situation particularly concerning is that Earth's energy imbalance—the difference between energy coming in and energy going out—is now larger than at any time in modern measurement history. We're essentially stockpiling heat, primarily in the oceans, which have absorbed about two-thirds of the total heating influence of all human-produced greenhouse gases. That stored heat will continue to affect our climate for decades to come. It's like we're running up a climate credit card bill that our grandchildren will still be paying off.

So What? The Impacts Beyond Sweaty Armpits

A warming planet doesn't just mean you'll save on winter heating bills (though your summer cooling costs will more than make up for it). The effects ripple through Earth's systems in ways more complex than the plot of a Christopher Nolan movie:

Global Temperature Patterns

Average temperatures have already increased beyond historical precedents, with many of these changes unprecedented in thousands, if not hundreds of thousands of years. That might sound trivial—"just a couple degrees, put on a lighter jacket, snowflake!"—but remember: when you have a fever of just 2°F, you feel terrible, post dramatic social media updates, and demand chicken soup as if it contained the elixir of life. Earth has a fever, and the only prescription isn't more cowbell—it's less carbon.

If we continue on our current emissions path, we're looking at warming of 3-4°C (5.4-7.2°F) by 2100. For reference, the difference between today's climate and the last ice age—when giant ice sheets covered much of North America and Europe—was about 5-6°C in the opposite direction. So we're talking about a planetary-scale transformation in the geologic equivalent of a lightning bolt. It's like remodeling your house with explosives instead of contractors—quicker, sure, but not exactly what you'd call "controlled."

Weather on Steroids

Climate change doesn't cause hurricanes, droughts, or floods—but it makes them stronger and more frequent. It's like adding steroids to the weather. That Category 3 hurricane might bulk up to a Category 4 or 5. That "once-in-a-century" flood might start showing up every few decades, like that relative who says they're "just passing through" but somehow stays for weeks.

As the atmosphere warms, extreme events that previously occurred once per century could happen every year by the end of this century, according to the IPCC. Think of it as the weather developing a more dramatic personality.

Rising Seas: Real Estate Agents Hate This One Trick

As the planet warms, two things happen to the oceans: the water itself expands (thermal expansion), and land ice melts, adding more water to the seas. Sea levels have already risen significantly, and the IPCC projects that without major investments in adaptation, many low-lying coastal cities and small islands would be exposed to escalating flood risks.

This isn't just a problem for beachfront property owners who might find themselves with unexpected indoor swimming pools. Globally, millions of people living on low-lying islands face this prospect, including inhabitants of Small Island Developing States (SIDS), densely populated deltas, coastal villages, and Arctic communities. Many of the world's great cities—New York, Miami, Shanghai, Mumbai, Amsterdam, Venice—face existential threats from sea level rise. Venice is already practicing by flooding regularly, but most other cities are woefully unprepared for their future as part-time submarines.

Ocean Acidification: The Other CO_2 Problem

Beyond warming the planet, carbon dioxide poses another threat when it dissolves in seawater: ocean acidification. About a quarter of the CO_2 we emit gets absorbed by the oceans, where it forms carbonic acid. The ocean has already absorbed enough carbon dioxide to lower its pH by 0.1 units, a 30% increase in acidity.

This increased acidity makes it harder for shellfish, corals, and certain plankton to build their calcium carbonate shells and skeletons. It's as if we're slowly dissolving the foundation of marine ecosystems. Since these organisms form the base of many marine food webs, the impacts could cascade through ocean ecosystems with potentially severe consequences for fisheries and the hundreds of millions of people who depend on them for protein.

The double whammy of warming and acidification puts coral reefs in particular peril. According to the IPCC's Special Report on the Ocean and Cryosphere, warming has increased the frequency of large-scale coral bleaching events, causing worldwide reef degradation since 1997-1998, with some reefs shifting to algal dominance. We've already lost about half the world's coral reefs, and without substantial emissions reductions and targeted conservation efforts, they could largely disappear by mid-century. It's like losing half the Amazon rainforest and shrugging it off because "it's underwater so who cares?"

Ecosystems: Nature's Great Migration

Plants and animals have evolved within specific climate conditions. When those conditions change rapidly, species must adapt, migrate, or die. Many are already shifting their ranges poleward or to higher elevations, like college students fleeing to the one air-conditioned room in the dorm during a heat wave.

Distributions of seagrass meadows and kelp forests are contracting at low-latitudes due to warming, with losses of 36-43% following heat waves in some areas. Warming has also increased the frequency of large-scale coral bleaching events. The timing of seasonal events is changing too—earlier spring blooms, earlier migrations, earlier budding of trees. This creates mismatches when interdependent species respond differently to climate cues. When birds arrive after the insects they feed on have already peaked, or pollinators emerge before or after the flowers they pollinate are in bloom, ecological relationships that have evolved over millennia fall out of sync. It's like showing up to a dinner party only to find everyone else already ate and left—awkward and unsatisfying for all involved.

Human Health: This Time It's Personal

Climate change isn't just an environmental issue—it's a public health emergency with more plot twists than a hospital drama. Rising temperatures directly increase heat-related illnesses and deaths, particularly among vulnerable populations like the elderly, outdoor workers, and those without access to air conditioning.

Changing climate conditions also affect the geographic range and seasonality of infectious diseases as warming water temperatures, acidification, and low oxygen levels create extreme marine events. Mosquitoes that carry dengue fever, Zika virus, and malaria are expanding their territories as previously inhospitable areas become warm enough for them to survive. Tick-borne diseases like Lyme disease are showing up in new regions and at higher altitudes. It's like pests are using climate change as their personal GPS upgrade.

Air quality suffers too. Higher temperatures accelerate the formation of ground-level ozone, a key component of smog that can trigger asthma attacks and cause other respiratory problems. Wildfires, becoming more frequent and intense due to climate change, spew particulate matter that can travel hundreds or thousands of miles, affecting air quality far from the fire itself. Nothing says "we're all connected" quite like breathing smoke from a forest fire happening three states away.

But Wait, I Heard That...

Let's address some common misconceptions faster than your uncle can share climate conspiracy theories at Thanksgiving dinner:

"The climate has always changed!"

Yes, and people have always died of natural causes, but that doesn't make murder okay. The issue isn't that climate changes—it's the unprecedented rate of change. Many of the changes observed in the climate are unprecedented in thousands, if not hundreds of thousands of years, according to the IPCC. It's the difference between easing your car to a stop over 100 feet versus hitting a brick wall. Both result in a stopped car, but one approach is considerably more traumatic for the passengers.

Natural climate changes, whether from orbital variations, solar output, or volcanic activity, typically unfold over thousands or tens of thousands of years. That gives ecosystems time to adapt and migrate. The current warming is happening over decades—essentially a geological instant—at a rate about 100 times faster than previous natural increases. It's like expecting evolution to happen at fast-forward speed, which works great in Pokémon but not so well in reality.

"Scientists don't agree on this."

Actually, they really do. Multiple studies have found that 97-99% of climate scientists agree that human-caused climate change is happening. The level of scientific consensus on climate change is similar to the consensus on the link between smoking and cancer.

If 97% of mechanics told you your brakes were failing, would you say "the jury's still out" and drive your car onto the freeway? If 97% of dentists told you that candy and soda cause cavities, would you start using Mountain Dew as mouthwash? The smart money says no.

"CO_2 is plant food!"

Plants do use CO_2 for photosynthesis, but that doesn't mean more is always better, just like water is essential for human life but drowning is still bad. Studies in natural settings show that increased CO_2 initially boosts plant growth in some species, but the benefit often diminishes as other factors (nutrients, water) become limiting.

Plus, climate changes caused by CO_2 increase (droughts, heat stress) can harm plant growth more than the extra "food" helps. It's like force-feeding someone chocolate cake while also cranking up the thermostat, removing their water supply, and occasionally setting their house on fire —the cake might be nice, but the overall package is problematic.

"It's cold today, so much for global warming!"

Weather isn't climate. Weather is what outfit you choose today; climate is your entire wardrobe. A cold day no more disproves climate change than a salad disproves obesity or finding a dollar disproves poverty.

This is like standing in a walk-in freezer and declaring "the inside of this freezer proves global warming is a hoax!" while ignoring the fact that the freezer itself is in a building that's gradually heating up, and the freezer's cooling system is working harder than ever to maintain temperature while consuming more electricity than it used to.

Solutions That Don't Require Moving to Mars

The good news? We already have many of the technologies and policies needed to address climate change. The challenge is scaling them up and implementing them quickly enough. It's like having all the ingredients for a cake but needing to bake it before the house burns down.

Clean Energy Revolution

Renewable energy, particularly solar and wind, has become dramatically cheaper over the past decade—so much so that in many regions, building new renewable capacity is now cheaper than operating existing fossil fuel plants. That's right: clean energy isn't just better for the planet; it's increasingly better for your wallet too.

Energy storage technologies are advancing rapidly to address the intermittency issue (the sun doesn't always shine, the wind doesn't always blow—much like my motivation to exercise). Meanwhile, energy efficiency measures continue to reduce demand, making the transition easier.

Solar energy deserves special attention for its remarkable cost decline—down roughly 90% since 2010. What was once a boutique option for the environmentally conscious wealthy is now often the cheapest form of new electricity generation. It's like organic food suddenly becoming cheaper than fast food—a plot twist nobody saw coming but everyone can benefit from.

Transportation Transformation

Electric vehicles (EVs) are moving from niche to mainstream, with major automakers committing to partial or complete transitions to electric fleets. As electricity generation becomes cleaner, EVs become an increasingly powerful climate solution.

But it's not just about swapping gas cars for electric ones. Reimagining transportation systems—expanding public transit, making cities more walkable and bikeable, optimizing freight—can reduce emissions while making our communities more livable. After all, nobody ever had a magical life experience while stuck in traffic, unless you count that time you found an old French fry that was still somehow edible.

Natural Climate Solutions

Sometimes the best technologies are the ones that evolved over millions of years. Protecting and restoring forests, wetlands, grasslands, and other ecosystems can remove carbon from the atmosphere through photosynthesis—the process through which trees and plants capture carbon and release oxygen.

Agricultural innovations like regenerative farming practices can turn farmland from a carbon source to a carbon sink while improving soil health and resilience. This approach recognizes that nature had pretty good carbon management systems before we came along with our fossil fuel addiction and Instagram food posts.

Policy and Social Change

Technology alone won't solve climate change. We need policies that accelerate the transition to clean energy, protect natural carbon sinks, and ensure the costs and benefits of climate action are distributed fairly.

Carbon pricing, whether through carbon taxes or cap-and-trade systems, helps incorporate the true cost of carbon pollution into economic decisions. Clean energy standards push utilities toward renewable sources. Building codes can dramatically reduce energy use in new construction. These policy tools are like the operating system for our climate solutions—without them, even the best technologies won't run effectively.

What You Can Do (Besides Panic)

Individual action matters, but systemic change matters more. Here's a guide to making your impact count without becoming that insufferable friend who makes everyone feel guilty for using a plastic straw:

Home Energy: The Low-Hanging Fruit

Revamp the way you heat, cool, and power your home by making it energy efficient. Through 2032, homeowners can use federal tax credits to cut energy-efficiency upgrade costs by up to 30% or $3,200 annually, as well as receiving a 30% tax credit for clean energy equipment.

- Get an energy audit and improve insulation—the clean energy equivalent of fixing a leaky bucket before adding more water
- Switch to LED lighting, which saves an average household more than $200/year compared to incandescent bulbs
- Choose energy-efficient appliances when replacing old ones
- Consider rooftop solar if your location and situation permit
- Look into community solar options if rooftop isn't feasible for you

This isn't just about reducing your carbon footprint—it's about saving money too. Energy efficiency improvements often pay for themselves many times over during their lifetimes, unlike that exercise equipment currently serving as an expensive clothes rack.

Transportation: Getting From A to B Without Destroying C(limate)

- Drive less when alternatives exist (walk, bike, public transit, carpool)
- If you're buying a vehicle, consider electric or highly efficient options
- Reduce air travel, and offset when you do fly
- Support policies that make low-carbon transportation more accessible in your community

Transportation represents the largest source of emissions in many developed countries, so changes here can have significant impact. Electric vehicles aren't just cleaner—they're also cheaper to maintain and operate over their lifetimes. Plus, they accelerate like they're auditioning for the next Fast & Furious movie.

Food Choices: Eating Like the Planet Depends on It (It Does)

- Reduce food waste by only buying what you need, composting food scraps, and storing food properly
- Eat lower on the food chain (more plants, less meat—especially beef)
- Support sustainable, local food systems when possible
- Try plant-based alternatives to animal products—they're getting tastier all the time!

The food system accounts for roughly a quarter of global greenhouse gas emissions, with animal agriculture being particularly carbon-intensive. Even small shifts toward plant-based eating can make a meaningful difference. And no, nobody's coming for your occasional bacon cheeseburger—we're talking about balance, not burger prohibition.

Use Your Voice and Vote: Democracy Is a Participation Sport

- Let elected officials know that climate action matters to you
- Support candidates with serious climate plans
- Talk about climate change—normalize the conversation and share solutions
- Join or support organizations working on climate issues

Political action is crucial because individual lifestyle changes alone won't get us where we need to go. We need policy changes, market transformations, and large-scale infrastructure investments. Your voice as a citizen and voter helps build the political will for these bigger shifts.

It's like being in a rowboat heading toward a waterfall—individual rowing techniques matter, but collectively deciding to change direction matters more.

Financial Action: Money Talks, Make Yours Say Something Worthwhile

Money talks, and where you put yours makes a difference. Banks use your deposits to finance various projects and industries—including, for many major banks, fossil fuel expansion. Between 2016 and 2021, the world's 60 largest banks financed fossil fuels to the tune of $4.6 trillion. That's trillion with a T, as in "Totally unnecessary continuation of climate disaster."

Consider switching to banks with stronger climate commitments, or at minimum, asking your current bank about their fossil fuel financing policies. Similar principles apply to investment accounts, retirement funds, and insurance companies—all of these financial institutions make decisions about where to put money, and those decisions have climate implications.

Divestment campaigns—which push institutions like universities, religious organizations, pension funds, and municipalities to remove investments from fossil fuel companies—have gained significant momentum in recent years. They aim not just to reduce financial support for the fossil fuel industry, but also to reduce its political legitimacy and social license to operate. It's like an intervention for a friend with an addiction, except the friend is a multi-trillion-dollar industry that's been gaslighting us for decades.

The Bottom Line: It's Bad, But Not Hopeless

Climate change is the ultimate collective action problem. No individual, company, or even country can solve it alone. But collectively, we have the knowledge, technology, and capacity to address it. It's like a global group project where nobody can afford to be the slacker who doesn't do their part.

The question isn't whether we can solve climate change—it's whether we will choose to do so quickly enough to avoid the worst impacts. As the IPCC states, strong and sustained reductions in emissions of carbon dioxide and other greenhouse gases would limit climate change, with benefits for air quality coming quickly, though it could take 20-30 years to see global temperatures stabilize.

Think of it this way: our actions today will echo through centuries. Future generations will look back at this moment as pivotal—when humanity faced its greatest collective challenge and either rose to the occasion or failed to act in time.

Let's make sure they tell the first story, not the second. Because let's be honest, post-apocalyptic fashions may look cool in movies, but in reality, they'd be itchy, impractical, and probably smell terrible.

As we move forward, it's important to balance urgency with hope. Climate scientist Katharine Hayhoe puts it well: "The most important thing you can do about climate change is talk about it." Breaking the silence, sharing solutions, and working collectively brings climate action from the abstract into our daily lives.

And despite the seriousness of the situation, there's genuine reason for optimism. According to the IPCC and the International Energy Agency, we already have affordable, reliable technologies that can put the peak in global emissions behind us and start the drive down to net zero. The transformation is underway—our job is to accelerate it.

In the coming chapters, we'll dive deeper into the specifics—the science behind extreme weather, the transformation of our energy systems, the role of natural ecosystems, and the political and economic dimensions of climate action. But the core message remains: this is the defining challenge of our time, we have the tools to address it, and our actions matter.

The greenhouse effect might not be a spa treatment, but with the right approach, we can turn down the heat and create a more sustainable, equitable, and resilient world. That's something worth fighting for— even if it doesn't come with complimentary cucumber water.

Chapter 2: Weather vs. Climate: Yes, They're Different

And no, your snowstorm doesn't disprove climate change.

"It's freezing outside! So much for global warming, right?"

Ah, the battle cry of uncles at Thanksgiving dinners across America. If you've ever heard this quip during a cold snap (or said it yourself—no judgment, we're all learning here), you've stumbled into the most persistent misconception in the climate conversation. It's the meteorological equivalent of thinking your diet is working great because you skipped dessert once, while ignoring the other 167 doughnuts you've eaten this month.

This confusion isn't just a matter of semantics. It's the cornerstone of countless misleading tweets, late-night talk show jokes, and political soundbites that spread faster than that stomach bug at your kid's daycare. So let's clear it up once and for all, shall we?

The Not-So-Secret Relationship Between Weather and Climate

Weather and climate are related, but they're not the same thing. Think of it as the difference between your mood and your personality.

Weather is what's happening in the atmosphere at a particular place and time. It's the temperature, humidity, precipitation, cloudiness, wind, and atmospheric pressure that you experience when you step outside and immediately regret your choice of footwear. Weather is inherently variable and can change rapidly—sometimes within the same day or even hour, much like your toddler's emotions.

Climate, on the other hand, is the average of weather over time—typically at least 30 years. The World Meteorological Organization defines climate as "the average weather conditions for a particular location and over a long period of time." It's the long-term pattern that emerges when you zoom out from the day-to-day fluctuations. If weather is the mood swings, climate is the underlying personality.

Here's a simple way to think about it:

- Weather is whether you need an umbrella today. Climate is whether you own an umbrella at all.
- Weather is what outfit you wear; climate is your entire wardrobe.
- Weather is a single Tinder date; climate is your entire relationship history (which, let's be honest, probably has some concerning patterns too).

Weather: Nature's Mood Swings

Weather is naturally variable—sometimes wildly so, like your significant other when they're hungry. This variability is influenced by factors such as:

1. **Daily cycles**: The sun rises and sets, causing temperature fluctuations. Revolutionary concept, I know.
2. **Seasonal cycles**: Earth's axial tilt gives us seasons, with their associated temperature patterns and retail marketing opportunities.
3. **Geography**: Mountains, oceans, and other geographical features affect local weather conditions. This is why Hawaii and North Dakota have slightly different winter experiences. And by

"slightly," I mean "completely and utterly different in every conceivable way."

4. **Atmospheric dynamics**: Air masses move and interact, creating fronts, pressure systems, and storms, which meteorologists point at enthusiastically on green screens while wearing suspiciously formal attire from the waist up.

This natural variability means that even as the planet warms overall, we still experience cold days, snowstorms, and record-breaking freezes in some places. Think of it like playing cards while riding up an escalator—you might draw low cards sometimes, but you're still moving upward overall. Or like gaining weight while still having days when your pants feel looser.

Climate change doesn't eliminate this variability; it shifts the entire distribution. The IPCC Sixth Assessment Report indicates that human-induced climate change is already affecting many weather and climate extremes across the globe, with increasingly strong evidence since the previous report. So we still have variability, but it happens around a new, warmer average. This means more extreme heat events, fewer extreme cold events (though they still occur), and a host of other changes to the patterns we've come to expect—like how shopping malls now put up Christmas decorations sometime around Labor Day.

Why Your Personal Experience Is Utterly Useless (For Climate Science)

Human memory and perception are incredibly limited tools for understanding climate change. We're hardwired to notice what's happening now and in our immediate surroundings, not gradual global trends over decades. It's like trying to watch grass grow—you won't see much happening in real-time, but leave for a month and suddenly your lawn looks like a jungle warfare training facility.

If you've ever shown a climate change graph to someone only to have them respond, "Well, it snowed a lot this winter where I live," you've witnessed this cognitive limitation in action. We naturally prioritize personal experience over abstract data, even when the data comes from thousands of measurements over many years. It's like dismissing the existence of world hunger because you just had a big lunch.

This tendency to overvalue personal experience leads to what scientists refer to as confusion between short-term weather events and long-term climate trends. When President Trump tweeted "What the hell is going on with Global Warming?" during cold weather, he was confusing weather with climate, a common misconception that scientists and media frequently need to correct. Survey data consistently show that people are more likely to express belief in climate change on hotter days and less likely on colder days, despite the fact that neither should logically affect one's understanding of global climate science. It's like basing your belief in gravity on whether you've tripped recently.

Weather 101: The Extremely Simplified Version

To understand how climate change affects weather, we need a quick primer on how weather works in the first place. Don't worry—we'll keep it simple enough that you could explain it to your cat, though whether your cat will care is an entirely different matter.

Weather begins with the sun warming Earth's surface unevenly. The equator receives more direct sunlight than the poles, creating temperature differences. These temperature differences create pressure differences in the atmosphere, which in turn drive winds and weather patterns. It's basically a giant, planet-sized game of hot and cold that creates air currents instead of awkward party games.

The rotation of Earth adds another layer of complexity, causing air and ocean currents to curve rather than flow directly from high to low pressure (the Coriolis effect). This gives us the familiar weather patterns we see on meteorological maps—cyclones and anticyclones spiraling across the globe like cosmic art projects.

Water is the wild card in this system. It moves between solid, liquid, and gas phases, storing and releasing energy as it changes state. This drives the water cycle and powers many weather phenomena, from rain showers to hurricanes. Water vapor is also a greenhouse gas, which means it traps heat in the atmosphere—it's like that friend who's great at parties but also somehow always starts drama.

The Big Picture: Weather Systems That Shape Our Lives

Individual weather events don't happen in isolation. They're part of larger-scale patterns and systems, kind of like how your questionable life choices aren't random but follow discernible patterns that your therapist is trying to help you recognize:

1. **Jet streams**: Fast-flowing air currents in the upper atmosphere that guide weather systems and separate air masses. The polar jet stream sits in the troposphere, characterized by a belt of wind that blows from west to east at speeds up to a couple of hundred miles per hour. They're like atmospheric highways, except the speed limits keep changing, and the road itself occasionally decides to meander wildly.

2. **El Niño/La Niña**: Periodic warming/cooling of the tropical Pacific Ocean that affects weather worldwide. Think of them as the world's weather DJs, mixing up climate patterns across continents. When El Niño drops a beat, farmers in Australia start sweating.

3. **Monsoons**: Seasonal wind pattern shifts that bring wet and dry seasons to many regions. They're like nature's version of feast or famine—either too much water or not enough, with very little in-between, kind of like your company's air conditioning.

4. **Ocean circulation**: Currents like the Gulf Stream that distribute heat globally. The planetary equivalent of those hot water pipes that keep your apartment habitable in winter, unless you live in that one unit where they never seem to work properly.

These larger patterns provide some predictability to weather and shape regional climates. For instance, the Mediterranean climate of coastal California is influenced by the cold California Current offshore and the positioning of the subtropical high-pressure zone. It's why Napa has great wine and why people in San Francisco always carry a light jacket, even in summer.

But these patterns are not fixed. They respond to changes in the climate system—and that's where climate change enters the picture, like that one friend who shows up to every party and somehow changes the entire vibe.

How Climate Change Is Giving Weather a Makeover

Climate change doesn't just make everything uniformly warmer (though overall warming is certainly happening). It fundamentally alters the energy balance of our planet, which affects weather patterns in complex ways. The IPCC report states that climate change is bringing multiple different changes in different regions, including changes to wetness, dryness, winds, snow, ice, coastal areas, and oceans. It's not just turning up the thermostat—it's more like rewiring your house while simultaneously rearranging the furniture and adopting several hyperactive pets.

More Energy = More Extreme Weather

The extra heat trapped by greenhouse gases doesn't just warm the air—it adds energy to the entire climate system. And more energy means more potential for extreme weather, like giving espresso to a toddler who's already bouncing off the walls. Here's how:

- **Warmer air holds more moisture**: For every 1°C (1.8°F) of warming, the atmosphere can hold about 7% more water vapor. This increases the potential for extreme precipitation events when conditions are right. It's like upgrading from a water pistol to a fire hose.

- **Increased evaporation**: As the lower atmosphere becomes warmer, evaporation rates increase, resulting in more moisture circulating throughout the troposphere. An observed consequence of higher water vapor concentrations is the increased frequency of intense precipitation events, mainly over land areas. Your garden and your skin moisturizer budget both know this phenomenon all too well.

- **More powerful storms**: Warmer ocean surface waters can intensify hurricanes and tropical storms, leading to more hazardous conditions as these storms make landfall. It's like switching from regular to premium fuel in your disaster engine.

The Jet Stream Identity Crisis

The Arctic and Antarctic, often called Earth's refrigerators, are warming faster than anywhere else on the planet. This polar amplification has profound effects on weather patterns globally. The northern polar jet stream is driven partly by the temperature contrast between masses of icy air over the North Pole and warmer air near the equator. It's like if one side of your hot tub suddenly got way hotter than the other—the nice, even circulation would start doing some weird things:

- **Weakening jet stream**: Jennifer Francis and Stephen Vavrus proposed that a reduction in the temperature gradient between the pole and the mid-latitudes actually weakens or slows the jet stream, causing it to become wavier. This links global warming directly to changes in the amplitude of the jet stream—and, by extension, its ability to pull cold air south. This can cause weather systems to move more slowly and persist longer in one place, increasing the risk of prolonged extreme events like heat waves, droughts, and flooding rains. It's like when the conveyor belt at the sushi restaurant slows down, and the same mediocre California roll keeps passing by your seat for 20 minutes.

- **Shifting storm tracks**: While the polar jet stream affects weather patterns in the mid-latitudes, through the United States to Europe, its southern equivalent (the subtropical jet stream) has more influence on areas between the equator and 30 degrees

North. Sometimes these two jet streams interact, layer on top of each other; and when they do, we tend to get really big storms. Places that built their infrastructure expecting certain weather patterns are now getting different ones, like showing up for a blind date and finding your grandparents' bridge club instead.

- **Ocean circulation changes**: Warming and freshwater input from melting ice are affecting major ocean currents like the Gulf Stream, with potential consequences for weather patterns across the North Atlantic region and beyond. Remember that movie "The Day After Tomorrow"? It was like that, except instead of happening in three dramatic days, it's a slow-motion transformation over decades that's harder to make into a compelling Jake Gyllenhaal vehicle.

It's Not the Same Everywhere: Regional Climate Impacts

Climate change is occurring in all regions, but the impacts vary significantly. For 1.5°C of global warming, there will be increasing heat waves, longer warm seasons, and shorter cold seasons. At 2°C of global warming, heat extremes would more often reach critical tolerance thresholds for agriculture and health. This regional complexity is one reason why "global warming" has largely been replaced by "climate change" in scientific discourse—it better captures the diverse range of impacts beyond just temperature increases.

For instance:

- **The Arctic**: The polar vortex's behavior has become more extreme as a result of climate change. As Paul Ullrich explains, warming of the Earth has led to the loss of Arctic sea ice, transforming a highly reflective icy surface to a dark absorptive surface. This change is warming higher latitudes and reducing the temperature difference between the warmer mid-latitude and polar regions. It's like the Arctic got put in the microwave while the rest of the planet is in a conventional oven.

- **The tropics**: Expanding toward the poles, bringing tropical climate conditions to previously temperate regions. Mosquitoes are essentially getting an expanded travel visa, ready to bring their family of diseases to visit your previously mosquito-free neighborhood.

- **The American West**: Experiencing more intense droughts and wildfires as precipitation patterns shift and temperatures rise. The real estate market's response? "Charred is the new beige! Open concept now includes missing walls and roof!"

- **The Northeast U.S. and Northern Europe**: The IPCC notes that heat extremes have become more frequent and more intense across most land regions since the 1950s, and all regions are affected. In terms of the impact on society, heatwaves are happening more regularly, are starting earlier and ending later. Temperatures over 40°C and even 50°C are becoming increasingly frequent in many parts of the world. Snow days are becoming either extinct or occurring in April, much to the confusion of school administrators everywhere.

Extreme Weather's Greatest Hits: The New Abnormal

According to the Intergovernmental Panel on Climate Change (IPCC)'s Sixth Assessment Report released in 2021, the human-caused rise in greenhouse gases has increased the frequency and intensity of extreme weather events. We're witnessing a shift from a stable climate with occasional extremes to a more volatile climate where extremes become increasingly common. It's like going from that one dramatic friend who occasionally causes a scene at dinner to an entire friend group composed entirely of reality TV contestants.

Heat Waves: The New Summer Blockbuster

The IPCC states it is virtually certain that "there has been increases in the intensity and duration of heatwaves and in the number of heatwave days at the global scale." They're becoming more frequent, more intense, and longer-lasting across most land regions, like that house guest who said they'd stay "just for the weekend" and is somehow still on your couch three weeks later.

Recent examples like the 2021 Pacific Northwest heat dome, which saw temperatures reach 121°F (49.6°C) in British Columbia, Canada, and the 2022 European heat waves that set all-time records across the continent, are consistent with climate model projections. It's as if Mother Nature looked at our climate models and said, "Hold my beer."

Heat waves are particularly dangerous because they put stress on human health, infrastructure, agriculture, and ecosystems. They're among the deadliest weather phenomena, especially in regions unaccustomed to extreme heat or lacking adequate cooling infrastructure. It turns out humans aren't actually designed to function in convection oven conditions—who knew?

When It Rains, It Really Pours

Rainfall intensity is expected to increase for most land areas, but the largest increases in dryness are expected in the Mediterranean, southwestern South America, and western North America. Globally, daily extreme precipitation events will likely intensify by about 7% for every 1 degree Celsius that global temperatures rise. It's like your body holding more water right before your period, except instead of bloating, entire cities get underwater.

The frequency and intensity of heavy rainfall events has increased since the 1950s and this is expected to continue. There are numerous examples of a month or even many months' worth of rainfall falling in a matter of hours or days, with devastating and deadly flooding—as seen in Africa, Asia, Europe, North and South America in the past few years. This seeming paradox—more intense rain in some places that are overall getting drier—is a hallmark of climate change. It's like getting both dehydrated and having to pee really badly at the same time.

The 2021 European floods, the 2022 Pakistan floods, and numerous recent flash flooding events in the United States all occurred in a climate altered by human activities. Attribution studies—analyses that determine the influence of climate change on specific weather events—have shown that many recent flooding disasters were made more likely or more severe by climate change. It's not that these events wouldn't have happened otherwise, but climate change loaded the dice, like that friend who definitely modified their Monopoly dice but insists they're just "really lucky."

Droughts: When "Dry" Becomes a Lifestyle

Climate change is intensifying the water cycle. This brings more intense rainfall and associated flooding, as well as more intense drought in many regions. Climate change is affecting rainfall patterns. In high latitudes, precipitation is likely to increase, while it is projected to decrease over large parts of the subtropics. It's like cranking up the dehumidifier in your basement, except the basement is the American Southwest and the dehumidifier runs on burning fossil fuels.

NASA scientists confirmed that major droughts and pluvials—periods of excessive precipitation and water storage on the landscape—have been occurring more often. The worldwide intensity of these extreme wet and dry events is closely linked to global warming. Warmer air causes more moisture to evaporate from Earth's surface during dry events. These climate-fueled droughts have wide-ranging impacts on agriculture, water supplies, wildfire risk, and ecosystems. Medieval knights had better water security than modern Las Vegas. Let that sink in—or rather, don't, because there's not enough water for things to sink.

Hurricanes on Steroids: Same Number, More Muscle

The relationship between climate change and tropical cyclones (hurricanes, typhoons, and cyclones depending on the ocean basin) is complex. Current research suggests that climate change is not necessarily increasing the total number of these storms, but it is affecting their characteristics in several ways:

1. **Intensity**: Warmer ocean surface waters can intensify hurricanes and tropical storms, leading to more hazardous conditions as these storms make landfall. It's like switching from regular cola to the extra-caffeinated variety—same drink, more punch.

2. **Rainfall**: IPCC research shows that extreme daily precipitation events are projected to intensify by about 7% for each 1°C of global warming. Hurricanes are becoming less "wind with some rain" and more "floating lakes that also have wind."

3. **Storm surge**: Higher sea levels mean that storm surges reach further inland and cause more damage. The ocean increasingly visits places that never invited it over.

4. **Movement**: Some evidence suggests tropical cyclones are moving more slowly and intensifying more rapidly than in the past. They're like that dinner guest who arrives quickly, then lingers way too long.

The devastating impacts of hurricanes like Harvey (2017), which dropped over 60 inches of rain on parts of Texas, and Dorian (2019), which stalled over the Bahamas with sustained winds of 185 mph, reflect these climate-influenced characteristics. These aren't your grandparents' hurricanes—they're hurricanes on performance enhancers.

Polar Vortex Gone Wild: When the Arctic Visits Texas

According to NOAA stratosphere expert Amy Butler, the polar vortex is a band of strong westerly winds that forms in the stratosphere between about 10 and 30 miles above the North Pole every winter. The winds enclose a large pool of extremely cold air. When the vortex weakens, shifts, or breaks down, the upheaval is often mirrored in the polar jet stream below, leading to cold air outbreaks in the mid-latitudes. It's like if the freezer section of your refrigerator started defrosting while still connected to the cooling system—it would mess up the entire appliance.

When the polar vortex is disturbed, it can slip off the North Pole, or even break into two or three separate rings. While the polar vortex itself is a natural phenomenon, a warmer Arctic with less predictable weather may make this kind of disturbance more likely, affecting weather patterns elsewhere. Some scientists propose that the rapidly warming Arctic is leading to a wavier, more meandering jet stream that allows cold Arctic air to spill farther south while warm air pushes farther north. This can lead to persistent, extreme weather conditions—whether heat waves, cold snaps, floods, or droughts. It's like removing the bouncer from a club—suddenly all kinds of characters are showing up in places they shouldn't be.

Weather Whiplash: Mother Nature's Mood Swings

The probability of compound events has likely increased in the past due to human-induced climate change and will likely continue to increase with further global warming. Concurrent heatwaves and droughts have become more frequent, and fire weather conditions (compound hot, dry and windy events) have become more probable in some regions. It's like going from a juice cleanse directly to an all-you-can-eat buffet, but for entire ecosystems.

Examples include the shift from extreme drought to extreme flooding in California in 2023, or the Texas freeze of February 2021 followed by abnormal warmth just weeks later. It's climate change giving us meteorological mood swings.

These rapid transitions are particularly challenging for infrastructure, agriculture, and ecosystems, which typically evolve or are designed based on historical climate patterns with gradually shifting seasons and more predictable extremes. When infrastructure and ecosystems face conditions they weren't designed for—especially in quick succession—the impacts can be devastating. It's like asking someone who's only ever driven a golf cart to suddenly pilot a Formula 1 race car and then immediately switch to a helicopter.

CSI: Climate—Attribution Science Connects the Dots

When extreme weather events occur, one of the first questions people ask is: "Was this caused by climate change?" This is like asking whether a specific home run was caused by a baseball player's steroid use. You can't attribute any single hit directly to performance-enhancing drugs, but you can demonstrate that they increase the likelihood and magnitude of home runs overall.

Attribution science concerns the identification of causes for changes in characteristics of the climate system (e.g., trends, single extreme events). Since 2013, there have been important new developments and knowledge advances on changes in weather and climate extremes, including human influence on individual extreme events, changes in droughts, tropical cyclones and compound events.

Attribution science has grown explosively since 2004, when a breakthrough study in Nature examined a heat wave in Europe. The IPCC's 2021 report reflects major advances in the science of attribution – understanding the role of climate change in intensifying specific weather and climate events such as extreme heat waves and heavy rainfall events. The results are often expressed as a change in likelihood —for example, "Climate change made this heat wave 30 times more likely" or "This extreme rainfall event was 20% more intense due to climate change." It's like comparing your odds of getting a parking ticket in a no-parking zone versus a legal spot—one scenario drastically increases your chances of an unpleasant outcome.

New attribution research published by NOAA in the Bulletin of the American Meteorological Society found that the extreme heat and drought experienced by California and Nevada from 2020-2021 was made six times more likely by climate change. Another study showed that a marine and terrestrial heat wave that affected Asia in 2021 is now 30 times more likely due to climate change. Some recent extreme heat events have been found to be virtually impossible without climate change. It's not just that climate change is loading the weather dice—for some events, it's replacing some of the dice faces with "extreme disaster."

The Snowball in the Senate: Public Confusion About Weather vs. Climate

Despite the scientific clarity on the difference between weather and climate, confusion persists in public discourse. This confusion is sometimes inadvertent and sometimes deliberately cultivated, like my "accidental" purchases every time there's a sale at my favorite store.

A classic example occurred in 2015 when a U.S. senator brought a snowball onto the Senate floor as "evidence" against global warming. This stunt perfectly exemplified the weather-climate confusion: using a short-term, localized weather event (snow in Washington, DC, in February) to try to refute a long-term, global climate trend. It's like bringing an ice cube to a discussion about global ocean temperature rise and saying "See? Cold!" while standing next to a boiling pot of water.

Trump's comments about cold weather and global warming received widespread derision from scientists and the media, with many articles pointing out that Trump was confusing short-term weather events with long-term climate, and that extreme cold weather still occurs in a warming world. Conversely, during heat waves, posts attributing the hot weather directly to climate change also spread, though these are at least directionally aligned with climate science even if they oversimplify the relationship.

News media can inadvertently contribute to the confusion too. Weather events make great news stories—they're immediate, visual, and impactful. Climate trends, in contrast, are gradual, statistical, and often invisible without specialized measurements. This creates a natural bias toward covering weather rather than climate, making it harder for the public to understand the larger context. It's like if your doctor only ever treated your symptoms but never mentioned you have an underlying chronic condition—you'd miss the big picture entirely.

So What? Making Sense of Weather in a Changing Climate

So where does this leave us? How do we make sense of weather events in a changing climate? Should we panic every time it's hot and celebrate every time it's cold? (Spoiler: no, that's exhausting and scientifically unsound.)

First, we need to recognize that no single weather event—not a heat wave, not a blizzard, not a hurricane—proves or disproves climate change. Climate is about patterns over time, not individual events. It's like how one salad doesn't make you healthy, and one pizza doesn't make you unhealthy—it's the pattern of eating that matters.

At the same time, we should understand that climate change is now influencing all weather. As climate scientist Kevin Trenberth famously put it, "All weather events are affected by climate change because the environment in which they occur is warmer and moister than it used to be."

This doesn't mean every weather event is more extreme—natural variability still plays a major role. But it does mean that climate change is loading the weather dice, making certain types of extreme events more likely and more intense. It's like playing poker with a deck that has extra aces—you might not get one every hand, but your chances are definitely higher than they should be.

When we experience unusual weather—whether it's uncomfortably hot, surprisingly cold, exceptionally wet, or remarkably dry—it's worth asking how climate change might be influencing the event, without jumping to simplified conclusions in either direction. Just don't be that person who posts "climate change???" every time it rains or "what global warming???" every time it snows. Nobody likes that person. Not even their mother.

What This Means for Your Future (Besides More Awkward Weather Small Talk)

Understanding the relationship between weather and climate isn't just an academic exercise or material for awkward small talk at parties. It has practical implications for how we plan and prepare for the future:

1. **Infrastructure planning**: Vulnerability to water-related impacts of climate change and extreme weather are already felt in all major sectors and are projected to intensify in the future, for

example, in agriculture, energy and industry, water for health and sanitation, water for urban sectors, and freshwater ecosystems. We need to build for the climate of the future, not the climate of the past. It's like buying clothes for a teenager based on their current size—by the time the clothes arrive, they won't fit anymore.

2. **Agricultural practices**: Three major shifts documented in precipitation patterns are: (a) some regions receiving more annual or seasonal precipitation and others less, (b) many regions have seen increased heavy precipitation, and many have seen either increases or decreases in dry spells, and (c) some regions have seen shifts towards heavier precipitation events separated by more prolonged dry spells. The old farmer's almanac is becoming about as reliable as your horoscope or that fortune cookie that promised you'd "come into great wealth" right before you got laid off.

3. **Public health systems**: Preparing for more frequent and intense heat waves, expanding disease vectors, and other climate-related health risks requires proactive planning and investment. It's like how we should have had pandemic planning before, well... you know. (Too soon?)

4. **Natural resource management**: Forests, watersheds, coastal areas, and other ecosystems face unprecedented stresses from changing weather patterns and extreme events. Adaptive management approaches are needed to maintain their health and the services they provide. These ecosystems are essentially

having a midlife crisis, but instead of buying a sports car, they're dying.

5. **Emergency response**: As extreme weather becomes more frequent and intense, emergency management systems need to evolve, with greater emphasis on prediction, preparation, and resilience rather than just response and recovery. FEMA might want to consider bulk ordering those "Days Without Disasters" signs with single-digit numbers.

Choose Your Own Future: Climate Projections

Looking ahead, we can expect weather patterns to continue changing as the climate warms further. Climate models project:

- More frequent and intense heat waves in most land regions. Heavy precipitation will generally become more frequent and more intense with additional global warming. At a global warming level of 4°C relative to the pre-industrial level, very rare heavy precipitation events would become more frequent and more intense than in the recent past, on the global scale and in all continents and regions. Your "it's not the heat, it's the humidity" complaints will evolve into "it's both the heat AND the humidity AND the fact that it's been this way for three straight weeks."

- The increase in the frequency of heavy precipitation events will be non-linear with more warming and will be higher for rarer events, with a likely doubling and tripling in the frequency of 10-year and 50-year events, respectively, compared to the recent past. This is like getting both dandruff AND an oily scalp—the worst of both worlds.

- An increase in near-surface atmospheric water holding capacity of about 7% per 1°C of warming explains a similar magnitude of intensification of heavy precipitation events that increases the severity of flood hazards when these extremes occur. The severity of very wet and very dry events increases in a warming climate. This will turn "waterfront property" into "distant memory of waterfront property."

- Continued intensification of tropical cyclones, with higher rainfall amounts and potentially higher wind speeds, giving us hurricanes that could blow Dorothy all the way to the Emerald City without any pit stops.

- More "compound events" where multiple hazards occur simultaneously or in rapid succession. The probability of compound events has likely increased in the past due to human-induced climate change and will likely continue to increase with further global warming. It'll be like that week when you got food poisoning, your car broke down, your water heater exploded, and your boss decided it was a great time for your performance review.

The degree of change depends largely on future greenhouse gas emissions. Under high-emission scenarios, the changes would be more dramatic; under low-emission scenarios, they would be less severe but still significant compared to historical norms. It's like choosing between a bad hangover and alcohol poisoning—neither is great, but one is definitely preferable.

The Weather Report You Actually Need

The next time someone points to a snowstorm as evidence against climate change or attributes a single hot day directly to global warming, you'll have the context to explain why weather and climate are different but related concepts. You can be that annoying person at the party who actually knows what they're talking about! (Though maybe find a more engaging way to share this information than saying "Well, actually..." Nobody likes that person either.)

Weather will always be variable—that's its nature. But the climate within which that variability occurs is changing due to human activities, primarily the burning of fossil fuels and the resulting greenhouse gas emissions. It's like a game where the rules are gradually changing while we play.

Understanding this relationship helps us interpret the world around us more accurately, plan more effectively for the future, and appreciate the urgency of addressing climate change through both mitigation (reducing emissions) and adaptation (preparing for changes already underway).

So check the weather forecast for tomorrow, by all means—but keep an eye on the climate trends that will shape our weather for decades to come. That way, when your grandkids ask why you didn't do more about climate change despite all the evidence, you'll at least be able to say, "I understood the difference between weather and climate, and I tried to explain it to others." Which is more than that senator with the snowball can say.

Chapter 3: Carbon Dioxide: Tiny Molecule, Global Drama Queen

Why it's not just "plant food," Karen.

Ladies and gentlemen, meet the star of our climate change show: carbon dioxide, or CO_2 if you're into the whole brevity thing. This unassuming molecule—just one carbon atom sandwiched between two oxygens like the world's tiniest Oreo—is currently reshaping our planet's future more dramatically than any social media influencer could ever dream of.

CO_2 is the Kardashian of atmospheric gases—it makes up less than 0.05% of the air we breathe, but it's responsible for about 80% of the climate drama. It's like that friend who doesn't take up much space in your group text but somehow dominates every conversation and changes all your plans.

So what's the deal with this molecular diva? Why does something so small and invisible wield so much power? And why can't we just ignore it and go back to consuming fossil fuels like we're at an all-you-can-burn buffet? Buckle up, climate curious friends—we're about to get cozy with carbon dioxide.

Carbon 101: The Life of the Planetary Party

Before we drag CO_2 for its climate crimes, let's acknowledge that carbon itself is pretty freaking amazing. As the fourth most abundant element in the universe, carbon is the chemical backbone of all life on Earth. It's in your DNA, your lunch, your yoga pants, and your favorite houseplant. Carbon's ability to form complex molecules by bonding with itself and other elements is what makes organic chemistry—and therefore life—possible.

Carbon is basically the LEGO of elements—endlessly versatile and ready to build almost anything. Without carbon, we'd be a planet of inorganic materials just sitting around not evolving, reproducing, or posting memes. Boring!

Carbon moves through our world in a natural cycle that's been operating for billions of years. Plants pull CO_2 from the air during photosynthesis, using the carbon to build their tissues and releasing oxygen as a waste product (thanks, plants!). Animals (including us) eat the plants (or eat the animals that eat the plants), using the carbon compounds for energy and structure, and exhaling CO_2 back into the atmosphere. When organisms die, decomposition returns more carbon to the atmosphere and soil. This elegant process essentially means "the Earth breathes too," cycling carbon between the atmosphere and living things in a delicate balance.

This elegant cycling of carbon—from atmosphere to living things and back again—kept atmospheric CO_2 levels relatively stable for thousands of years. It's like a perfectly balanced checking account: deposits and withdrawals in harmony.

Carbon Dioxide: The Greenhouse Gas That Puts the "Green" in Greenhouse

As a greenhouse gas, carbon dioxide has one job: letting sunlight in while making it harder for heat to escape. Sunlight passes through the atmosphere relatively unimpeded, warming Earth's surface. That warmth radiates back upward as infrared energy, but CO_2 and other greenhouse gases absorb some of this outgoing heat and re-emit it in all directions—including back toward Earth's surface (NOAA Climate.gov).

This greenhouse effect is actually essential—without it, Earth would be a frozen wasteland with an average temperature of about -18°C (0°F). We'd all be popsicles with existential dread. But as with most good things (chocolate, Netflix, validation from strangers on the internet), there can be too much of it according to NOAA Climate.gov's explanation of how carbon dioxide traps heat as Earth's most important greenhouse gas.

Carbon dioxide isn't the most powerful greenhouse gas per molecule—methane and water vapor are both stronger heat trappers. But CO_2 gets the climate change spotlight for three key reasons:

1. **Persistence**: CO_2 hangs around in the atmosphere for a long time—hundreds to thousands of years. It's like that houseguest who says they'll stay "just a few days" but is somehow still on your couch three years later.

2. **Abundance**: We're pumping out about 36 billion tons of CO_2 annually, far more than other greenhouse gases. It's like comparing someone who occasionally tells a white lie (other greenhouse gases) to a pathological liar who makes up new stories every five minutes (CO_2).

3. **Cumulative effect**: Because of its persistence, CO_2 accumulates in the atmosphere. Each year's emissions pile on top of previous years', like layers in a particularly depressing geological cake.

A Brief History of Atmospheric CO_2: From Prehistoric Balance to Modern Mayhem

For much of human civilization, atmospheric CO_2 concentration hovered around 280 parts per million (ppm). That's 280 CO_2 molecules per million molecules of air—a tiny fraction, but enough to keep our planet comfortably habitable through the development of agriculture, the rise and fall of empires, and the invention of both cheese and smartphones (IPCC AR6).

Then came the Industrial Revolution, when humans discovered they could burn ancient, carbon-rich fuels like coal, oil, and natural gas to power exciting new technologies. These fossil fuels—formed from plants and tiny marine organisms that died millions of years ago—had been safely storing their carbon underground until we decided to dig them up and set them on fire. It's like taking all the carbon that plants worked so hard to sequester over millions of years and yelling "SURPRISE! WE'RE PUTTING IT BACK!" (NASA).

The result? Atmospheric CO_2 has shot up by 50% since the industrial revolution—meaning the amount of CO_2 is now 150% of its value in 1750 (NASA). This rate of increase is unprecedented in Earth's geological record outside of cataclysmic events like asteroid impacts or massive volcanic eruptions. It's the atmospheric equivalent of going from a sensible weekly budget to maxing out all your credit cards in a Vegas weekend.

Ice core samples from Antarctica, which contain tiny air bubbles trapped in ice laid down over hundreds of thousands of years, show that current CO_2 levels are higher than at any point in at least the last 800,000 years —possibly even the last 4 million years (NOAA Climate.gov). To find a period with comparable CO_2 levels, we'd need to go back to the Pliocene epoch, 3-5 million years ago, when sea levels were 5-40 meters (16-131 feet) higher than today and global temperatures were 2-3°C warmer (IPCC AR6). Hope you didn't just invest in beachfront property.

Carbon Budget: Earth's Maxed-Out Credit Card

Climate scientists talk about Earth's "carbon budget"—the finite amount of carbon we can emit while having a reasonable chance of limiting warming to specific targets (like 1.5°C or 2°C above pre-industrial levels) (IPCC). Think of it as a collective credit card with a strict limit. Once we hit that limit, we're in for some serious planetary consequences—and unlike your personal credit, we can't declare bankruptcy and start over.

How much is left in our carbon budget? For a 50% chance of limiting warming to 1.5°C, we have about 580 billion tons of CO_2 left to emit from 2018—which at current rates would be used up in just a few decades (IPCC). The IPCC's Sixth Assessment Report updated these numbers, noting that for a 67% chance of staying below 1.5°C, the carbon budget is extremely tight, requiring deep and rapid cuts in emissions (IPCC AR6).

According to the Mercator Research Institute, if we stay at current emission levels, we would use up our 1.5°C carbon budget in less than six years (MCC-Berlin). It's like having just enough money for rent but still needing to buy groceries, medicine, and pay utilities—something's gotta give.

The budget concept is useful because it makes clear that every ton of CO_2 counts, and that delaying action means we'll need steeper cuts later

—or will exceed our temperature targets (Carbon Brief). It's like waiting until the day before a term paper is due to start writing: the longer you procrastinate, the more painful the all-nighter becomes.

Where All This CO_2 is Coming From: Pointing Fingers Productively

So who invited this CO_2 party crasher? Where is all this extra carbon dioxide coming from? Let's break it down:

Fossil Fuels: The Carbon Time Capsules We Should Have Left Buried

Burning fossil fuels accounts for about 75% of the CO_2 emissions added to the atmosphere since the Industrial Revolution (IPCC AR6). Every time we burn coal, oil, or natural gas for electricity, transportation, heating, or industrial processes, we're releasing carbon that was safely stored away for millions of years.

Different fossil fuels release different amounts of CO_2 per unit of energy produced (EPA):

- Coal is the dirtiest, releasing about 2.2 pounds of CO_2 per kilowatt-hour
- Oil releases about 1.6 pounds per kilowatt-hour
- Natural gas is the "cleanest" fossil fuel, releasing about 1.1 pounds per kilowatt-hour

But "cleanest fossil fuel" is like "most hygienic garbage dump"—it's still a problem in the grand scheme of things.

Deforestation: Chopping Down Carbon Sponges

Trees and forests are basically carbon dioxide vacuum cleaners—they suck CO_2 from the air and store it in their wood, roots, and the soil around them. When we cut down forests for agriculture, timber, or development, we're not just losing future carbon absorption capacity; we're often releasing carbon that was already stored. Deforestation and land-use changes account for about 10-15% of global CO_2 emissions (IPCC).

The Amazon rainforest alone contains billions of tons of carbon—equivalent to several years of global fossil fuel emissions (World Resources Institute). As human activities chip away at this and other forests, we're eroding a critical carbon sink. It's like firing the cleaning crew while simultaneously throwing an even messier party.

Cement Production: The Overlooked Carbon Criminal

The process of making cement—the key ingredient in concrete—is responsible for about 8% of global CO_2 emissions (IPCC). This makes the cement industry the third-largest industrial source of carbon pollution in the world.

The issue is twofold: producing cement requires heating limestone to extremely high temperatures (typically using fossil fuels), and the chemical process itself releases CO_2 when calcium carbonate breaks down into calcium oxide and carbon dioxide (Carbon Brief). It's a double carbon whammy that gets overlooked in many climate discussions, despite the fact that we produce about 4 billion tons of cement annually.

Next time you admire a sleek concrete building, remember that its climate impact might be as solid as its foundation.

"But CO_2 is Plant Food!" And Other Misleading Arguments

Whenever CO_2's climate impact comes up, someone inevitably responds with "but plants need carbon dioxide for photosynthesis!" This is technically true in the same way that "humans need water to survive" is true, but it doesn't mean flooding your house is a good idea.

Plants do indeed use CO_2 for photosynthesis, and laboratory studies show that many plants grow faster under elevated CO_2 conditions (a phenomenon called the "CO_2 fertilization effect"). But—and this is a big but—there are several important caveats:

1. **Diminishing returns**: Most plants don't continue responding to ever-higher CO_2 levels indefinitely. They typically reach a saturation point where other factors (like water, nutrients, or light) become limiting.

2. **Reduced nutrition**: Higher CO_2 often leads to less nutritious plants with lower concentrations of protein, zinc, and iron—a serious concern for food security in a world where billions already face nutrient deficiencies.

3. **It's not just about CO_2**: Climate change brings higher temperatures, altered precipitation patterns, and more extreme weather events—all of which can harm plant growth and agricultural productivity despite the fertilization effect.

4. **Weeds love CO_2 too**: Many invasive species and agricultural weeds respond more positively to elevated CO_2 than crop plants

do. Poison ivy, for instance, grows larger and produces more irritating oils under high-CO_2 conditions. Great news for masochists, bad news for everyone else.

5. **Ecosystem impacts**: Even if some plants benefit from higher CO_2, the disruption to ecosystems, pollinators, and the timing of seasonal events can outweigh the positives.

So yes, plants use CO_2. But using that fact to dismiss climate concerns is like saying water is good for crops, therefore floods are beneficial. Context matters, people!

The Case of the Missing Carbon: Where Does It All Go?
Here's a puzzler: humans emit about 35 billion tons of CO_2 annually, but only about half of that actually stays in the atmosphere. So where does the rest go? Are we getting a 50% discount on our carbon crimes?

Not exactly. We're just lucky that Earth has some fantastic natural carbon sinks:

1. **Oceans**: The world's oceans absorb about 25% of the CO_2 we emit. The carbon dioxide dissolves in seawater, forming carbonic acid in a process called ocean acidification—which is great for reducing atmospheric CO_2 but terrible for coral reefs, shellfish, and marine ecosystems. It's like treating a headache with a remedy that causes stomach ulcers.

2. **Forests and vegetation**: Plants and soils soak up another 25% of our emissions through photosynthesis and carbon sequestration in biomass and soil organic matter. As long as the

plants don't die and decompose (or burn), that carbon stays out of the atmosphere.

3. **The atmosphere**: The remaining 50% stays airborne, accumulating year after year and driving climate change. This is the carbon we actually measure when we talk about atmospheric CO_2 concentrations.

The bad news? These natural sinks show signs of strain. Warmer oceans can hold less dissolved gas, meaning the ocean sink may weaken over time. And climate change itself threatens forests through increased drought, wildfire, and pest outbreaks. If these sinks falter, a larger fraction of our emissions would remain in the atmosphere, accelerating warming in a vicious feedback loop.

It's like relying on your rich aunt to bail you out of credit card debt, then discovering she's cutting back on her own spending. Eventually, the bill comes due.

The Carbon Dioxide-Temperature Connection: A 66-Million-Year Reality Check

The link between CO_2 and global temperature isn't just theoretical—it's backed by extensive historical evidence. Ice cores, sediment layers, fossil records, and other paleoclimate indicators all tell the same story: when atmospheric CO_2 rises, global temperatures follow suit.

The reverse is also true—periods of lower CO_2 correspond with cooler climates. The ice ages that repeatedly gripped Earth over the past million years coincided with CO_2 levels around 180-200 ppm, while warmer interglacial periods saw levels around 280 ppm.

Go back further and the pattern holds. During the Paleocene-Eocene Thermal Maximum about 56 million years ago, a natural (but much slower) release of greenhouse gases raised global temperatures by 5-8°C, causing major ecosystem disruptions and a significant extinction event. During the Cretaceous period (145-66 million years ago), CO_2 levels were several times higher than today, and dinosaurs roamed a planet with no polar ice caps and sea levels 100 meters higher than present.

This consistent relationship between CO_2 and temperature across vast timescales underscores a simple truth: greenhouse gases drive Earth's thermostat. The difference today isn't the mechanism—it's the unprecedented speed of change and the fact that human civilization developed during an unusually stable climate period that we're now disrupting.

Carbon's Fingerprints: How We Know Humans Are Responsible

At this point, you might be wondering: how do we know for sure that humans are causing the CO_2 increase? Maybe it's volcanoes? Or the sun? Or just natural cycles? (Spoiler: it's definitely us.)

Scientists have multiple lines of evidence pointing to human activities as the primary driver of rising CO_2 levels:

1. **Carbon isotope analysis**: Fossil fuel carbon has a distinct isotopic signature because it comes from ancient plants that preferentially absorbed carbon-12 over carbon-13. As we burn more fossil fuels, atmospheric CO_2 shows an increasing proportion of carbon-12—a smoking gun linking the increase to fossil sources rather than natural ones.

2. **Oxygen depletion**: When fossil fuels burn, they consume oxygen. Precise measurements show atmospheric oxygen declining in exact proportion to the amount expected from fossil fuel combustion.

3. **Volume calculations**: Human activities release about 35 billion tons of CO_2 annually. The total amount of carbon dioxide in the atmosphere has increased along with human emissions since the start of the Industrial Revolution. When the math is done on forest clearing, cement production, and fossil fuel use, the numbers match observed atmospheric increases (after accounting for natural sinks).

4. **Ruling out alternatives**: Volcanic emissions amount to less than 1% of human CO_2 output according to NRDC analysis of IPCC reports. Solar activity has been flat or declining while temperatures rise. Natural orbital cycles would actually be pushing us toward cooling right now. No natural explanation can account for the observed changes.

It's like arriving at a crime scene to find the suspect's fingerprints, DNA, a signed confession, and video footage of the act. The case against human-caused CO_2 increase isn't just beyond reasonable doubt—it's beyond any doubt whatsoever.

Why Small Numbers Matter: The Powerful Minority of CO_2

One common misconception is that CO_2 can't possibly matter much because it's such a small percentage of the atmosphere—currently about 0.042% or 420 parts per million. "How could something so tiny have such a big effect?" skeptics ask.

This line of thinking confuses abundance with impact. Many things have outsized effects despite small concentrations:

- A blood alcohol concentration of 0.08% (800 ppm) is enough to make you legally drunk in most countries
- Arsenic is toxic at just 300 ppb (parts per billion) in drinking water
- Many medications are effective at doses measured in micrograms
- Just a few drops of hot sauce can transform an entire pot of chili

In the case of CO_2, what matters isn't its overall abundance but its specific physical properties—namely, its ability to absorb and re-emit infrared radiation (heat) at wavelengths that would otherwise escape to space. It's like adding a lightweight but extremely effective blanket to your bed—it doesn't take up much space, but it dramatically affects how much heat is retained.

Moreover, small changes in average global temperature translate to much larger regional variations and significant impacts on weather patterns, ecosystems, and human systems. The difference between pre-industrial temperatures and today is only about 1.1°C (2°F), but that's enough to melt trillions of tons of ice, raise sea levels, and increase extreme weather events worldwide.

For perspective, the difference between today's climate and the depths of the last ice age—when giant ice sheets covered much of North America and Europe—was only about 5-6°C. Small temperature changes are a big deal for a planet as finely balanced as ours.

Feedback Loops: When Carbon Problems Compound

As if human emissions weren't concerning enough, Earth's climate system includes several feedback mechanisms that can amplify (or in some cases, dampen) the initial warming caused by CO_2 (IPCC AR6). These feedback loops make the climate response more complex—and potentially more dangerous—than a simple linear relationship between emissions and temperature would suggest.

Some major feedback mechanisms include:

1. **Water vapor feedback (positive)**: As the atmosphere warms, it holds more water vapor—itself a powerful greenhouse gas. More water vapor leads to more warming, which allows for more water vapor, creating an ongoing, continually amplified cycle (Climate Reality Project). This is the largest positive feedback in the climate system.

2. **Ice-albedo feedback (positive)**: White snow and ice reflect 84% of incoming solar radiation. But once we remove our icy protective shield, water reflects as low as 5% of solar radiation (Earth How). This reduction in reflectivity, or albedo, when ice melts contributes to a positive feedback loop, amplifying global warming.

3. **Cloud feedbacks (mixed)**: Changes in cloud cover and type can either amplify or reduce warming, depending on the cloud characteristics. Low, thick clouds mainly reflect sunlight (cooling), while high, thin clouds primarily trap heat (warming). The net effect of cloud changes remains an active area of research (IPCC AR6).

4. **Carbon cycle feedbacks (mostly positive)**: Several climate-sensitive carbon reservoirs could release additional greenhouse gases as the planet warms:

- Thawing permafrost could release CO_2 and methane from previously frozen organic matter, representing approximately twice the amount of carbon currently in Earth's atmosphere (PNAS)
- Warming oceans can't absorb as much CO_2 and may release some of what they've already taken up (MIT Climate Portal)
- Increased wildfires and drought-stressed forests may release stored carbon rather than absorbing it (Climate Reality Project)
- Warming wetlands and rice paddies could emit more methane (Dean et al.)

These feedbacks help explain why past climate changes often proceeded in a non-linear fashion, with relatively small initial triggers sometimes leading to substantial changes as various feedback mechanisms kicked in (Borrowed Earth Project). They're like compound interest on your climate change debt—the longer you wait to address the problem, the more the consequences compound.

Tipping Points: When Carbon Pushes Too Far

Perhaps the most concerning aspect of our carbon experiment isn't the gradual, predictable warming but the risk of crossing critical thresholds or "tipping points"—where a relatively small additional change triggers a large, often irreversible shift in a major Earth system (IPCC AR6).

Potential tipping elements in the climate system include:

1. **Arctic sea ice**: As reflective ice is replaced by heat-absorbing open water, warming accelerates, potentially leading to ice-free summers and fundamentally altered Arctic ecosystems (Borrowed Earth Project).

2. **Greenland and West Antarctic ice sheets**: These massive ice formations may have thresholds beyond which their disintegration becomes unstoppable, ultimately raising sea levels by meters (though over centuries rather than decades) (IPCC).

3. **Amazon rainforest**: In the Amazon, we've already lost one-fifth of the rainforest to climate change and human-caused burning. Scientists warn that losing another fifth would trigger a phenomenon known as "dieback," where the forest dries beyond human rescue, inviting more wildfires and releasing more carbon (Climate Reality Project).

4. **Atlantic Meridional Overturning Circulation (AMOC)**: This ocean current system, which includes the Gulf Stream, could weaken or even shut down due to freshwater input from melting ice, dramatically altering regional climate patterns, especially in Europe (Borrowed Earth Project).

5. **Permafrost and methane hydrates**: Vast stores of frozen greenhouse gases could be released as warming continues. Researchers estimate there may be as much as 1,600 billion tons of carbon stored in permafrost, potentially triggering a runaway greenhouse effect (PNAS).

The exact warming thresholds for these tipping points remain uncertain, but the risk increases substantially beyond 1.5-2°C of warming—which is why international climate agreements focus on these temperature targets (IPCC). It's like driving on a foggy mountain road—you know there's a cliff edge somewhere ahead, but you can't see exactly where it is, so prudence suggests slowing down rather than accelerating.

The Carbon Solution Set: From Individual Actions to System Change

So what do we do about our carbon dioxide problem? The good news is that we have many potential solutions. The challenging news is that we need to implement them at unprecedented speed and scale (IPCC). Let's explore the carbon solution set from smallest to largest scale:

Individual Actions: Small But Meaningful

Personal choices can reduce your carbon footprint while sending market signals that support broader change (Carbon Brief):

- **Transportation**: Walk, bike, use public transit, or drive electric vehicles. Reduce air travel or purchase high-quality carbon offsets when flying (IPCC AR6).

- **Home energy**: Improve insulation, switch to heat pumps and induction cooking, install solar panels if possible, and choose renewable electricity providers (WEF).

- **Food choices**: Reduce meat consumption (especially beef), minimize food waste, and choose locally grown, seasonal foods where practical (IPCC).

- **Consumption patterns**: Buy less stuff, choose durable products, repair instead of replace, and support companies with genuine climate commitments (Carbon Brief).

These actions matter—collectively, they can drive market shifts and policy change. But they're not sufficient on their own. It's like bringing your own bags to the grocery store while the store continues using non-recyclable packaging for everything inside.

Corporate Actions: Following the Money
Businesses are responsible for a large portion of global emissions and have significant power to reduce them (WEF):

- **Clean energy procurement**: Companies like Google, Apple, and Microsoft are purchasing 100% renewable electricity, driving demand for new clean energy projects (Climate Champions).

- **Supply chain decarbonization**: Major corporations are pressuring suppliers to reduce emissions, creating ripple effects throughout the global economy (IPCC).

- **Product redesign**: From electric vehicles to energy-efficient appliances, companies are reimagining products for a low-carbon future (Carbon Brief).

- **Internal carbon pricing**: Many corporations now apply an internal price on carbon when making investment decisions, steering capital toward lower-emission options (WEF).

Corporate action is necessary but subject to the constraints of profit motives and shareholder expectations. Some leading companies are genuinely transforming their business models, while others engage in greenwashing—making minimal changes while marketing themselves as environmental champions (IPCC).

Policy Solutions: Changing the Rules of the Game

Government policies can accelerate the transition to a low-carbon economy by changing economic incentives and investment patterns (IPCC AR6):

- **Carbon pricing**: Whether through carbon taxes or cap-and-trade systems, putting a price on carbon pollution helps internalize its true costs and drives investment toward cleaner alternatives (Carbon Brief).

- **Clean energy standards**: Requirements for utilities to generate increasing percentages of electricity from zero-carbon sources have proven effective in many jurisdictions (WEF).

- **Vehicle emissions standards**: Regulations requiring improved fuel efficiency or increasing sales of zero-emission vehicles drive transportation sector decarbonization (IPCC).

- **Building codes**: Updated standards for new construction can dramatically reduce energy use and enable electrification of heating and cooking (Carbon Brief).

- **Research and development funding**: Government investment in clean energy innovation has been crucial for technologies like solar PV and batteries, whose costs have fallen dramatically as a result (WEF).

Policies need to balance effectiveness with equity, ensuring that the clean energy transition doesn't burden vulnerable populations and that its benefits are widely shared (IPCC).

Technological Solutions: Innovation to the Rescue?

A range of technologies, both existing and emerging, will be crucial for addressing the carbon dioxide challenge (IPCC AR6):

The Clean Energy Revolution

Limiting global warming will require major transitions in the energy sector, including substantial reduction in fossil fuel use, widespread electrification, improved energy efficiency, and use of alternative fuels like hydrogen (IPCC).

Renewable energy sources like solar and wind have seen dramatic cost declines—90% for solar PV and 70% for wind since 2010 (IPCC). In many markets, they're now the cheapest forms of new electricity generation, even without subsidies. This economic transformation is driving a surge in renewable deployment worldwide (WEF).

Energy storage technologies, from batteries to pumped hydro, are addressing the intermittency challenge of renewables, enabling higher penetrations of variable sources on the grid. Meanwhile, smart grid technologies are making electricity systems more flexible and resilient (Carbon Brief).

Electrify Everything (Almost)

The IPCC's Sixth Assessment Report notes that cities can only achieve net-zero emissions through "deep decarbonisation and systemic transformation", which includes widespread electrification (Carbon Brief).

Electrification of end uses—from vehicles to home heating to industrial processes—allows them to be powered by increasingly clean electricity rather than on-site fossil fuel combustion (IPCC). Electric vehicles, for instance, are more efficient than internal combustion engines even when powered by a fossil-fuel grid, and their carbon footprint improves automatically as the grid gets cleaner (WEF).

Heat pumps for space and water heating can replace gas furnaces and water heaters, offering greater efficiency while eliminating point-of-use emissions (IPCC AR6). Induction cooktops provide a cleaner, safer alternative to gas stoves. It's a transition from burning stuff to using increasingly clean electrons—like upgrading from a campfire to a microwave, but for your entire life.

Carbon Capture: Cleaning Up Our Mess

While preventing emissions is the top priority, we've already put so much carbon dioxide into the atmosphere that we also need to consider removing some of it (IPCC):

- **Natural approaches**: Reforestation, improved agricultural practices, and ecosystem restoration can enhance natural carbon sinks. These "nature-based solutions" often bring co-benefits for biodiversity, water quality, and community resilience (WRI).

- **Technological approaches**: Direct air capture technologies can pull CO_2 directly from the atmosphere, though they remain expensive and energy-intensive. Carbon capture at industrial facilities and power plants can prevent emissions from reaching the atmosphere in the first place (IPCC).

- **Enhanced natural processes**: Methods like enhanced rock weathering can accelerate natural geological processes that remove CO_2 from the atmosphere (WEF).

Carbon removal isn't a substitute for emissions reductions—it's both more expensive and less certain. But given our depleted carbon budget, all pathways that limit warming to 1.5°C with no or limited overshoot depend on some quantity of carbon removal (WRI). Think of it as hiring a cleaning crew after a wild party—necessary, but not as good as preventing the mess in the first place.

The Political Economy of Carbon: Who Pays, Who Profits

The technical solutions to our carbon dioxide problem exist and continue to improve. The real challenges are political, economic, and social (IPCC):

Fossil Fuel Interests: Carbon's Powerful Defenders

The companies and countries that profit from fossil fuels have strong incentives to delay the clean energy transition. The fossil fuel industry spends hundreds of millions of dollars annually on lobbying and has historically funded climate misinformation campaigns (NRDC).

This resistance is understandable from a narrow economic perspective —global fossil fuel reserves represent trillions of dollars of potential revenue. If climate policies restrict burning these fuels, those assets become "stranded"—worthless except as museum pieces (Carbon Brief).

It's like owning a warehouse full of VHS tapes when DVDs were invented, except the VHS tapes are causing climate change.

Jobs and Communities: The Human Side of Transition

The clean energy transition is creating millions of new jobs globally, but these aren't always in the same locations as fossil fuel jobs. Communities built around coal mining, oil drilling, or gas extraction face genuine economic challenges as these industries decline (IPCC).

Just transition policies aim to support workers and communities through this shift, providing training, investment, and economic diversification (WEF). Without such measures, the transition risks leaving people behind and generating political backlash. It's like replacing a problematic restaurant in your neighborhood—great for public health, but you still need to make sure the workers can find new jobs.

Global Equity: The North-South Divide

Historically, developed countries have emitted far more carbon dioxide per capita than developing nations, yet climate impacts fall disproportionately on poorer countries with fewer resources for adaptation (IPCC AR6). This creates thorny questions of justice and responsibility in international climate negotiations.

Developing countries reasonably argue that they should have space to grow their economies and improve living standards, while developed nations should take the lead in emissions reductions and provide financial and technological support (WRI). Balancing these concerns remains one of the central challenges of global climate governance.

Carbon Timeline: The Race Against the Clock

The science is clear: to limit global warming to around 1.5°C, global greenhouse gas emissions need to peak before 2025 and be reduced by 43% by 2030, with methane also reduced by about a third (IPCC).

That's an enormously challenging timeline that requires transforming our energy systems, transportation, buildings, and industry at unprecedented speed (WEF). Every year of delay makes the necessary transition steeper and more disruptive (IPCC).

The good news is that many of the necessary changes are accelerating. Momentum is building for the net-zero transition, with the world spending $755 billion on low-carbon energy technologies in 2021, a 27% increase from the previous year (Climate Champions). But they're still not scaling fast enough to meet climate targets without stronger policies and investment (Carbon Brief).

The carbon timeline isn't just about avoiding the worst impacts of climate change—it's also about capturing the economic benefits of leading the clean energy transition rather than following it. Countries and companies that move aggressively toward decarbonization are positioning themselves for leadership in the industries of the future (IPCC).

The Psychological Challenge: Why Carbon Dioxide Is Hard to Care About

Beyond the technical, economic, and political challenges, carbon dioxide presents a psychological challenge. It's colorless, odorless, invisible, and —at current concentrations—directly harmless to breathe. Its effects are gradual, cumulative, and globally distributed rather than immediate and local (Climate Reality Project).

These characteristics make CO_2 the perfect villain for our cognitive blind spots:

- **Abstractness**: Unlike other forms of pollution that can be seen, smelled, or directly felt, carbon dioxide's impact requires understanding complex causal chains that our brains didn't evolve to track easily (MIT Climate Portal).

- **Temporal distance**: The most severe consequences unfold over decades rather than days or weeks, while humans tend to focus on immediate threats and rewards (IPCC).

- **Spatial distance**: Emissions in one location affect the entire planet, diluting the feedback between actions and consequences that normally guides our behavior (Carbon Brief).

- **Collective action problem**: No individual, company, or even country can solve climate change alone, creating a classic "tragedy of the commons" where rational individual behavior leads to collective harm (IPCC AR6).

Overcoming these psychological barriers requires making climate change more concrete, immediate, and personal—connecting it to values people already hold and demonstrating how solutions can improve lives in tangible ways (Climate Reality Project). It's like how public health campaigns succeeded against smoking not by focusing on abstract statistics but by telling compelling stories and connecting to deeply held values like protecting children.

Carbon Dioxide in Context: The Elephant in an Increasingly Hot Room

To wrap up our carbon dioxide deep dive, let's put this molecule in its proper context:

1. It's the main driver of climate change, responsible for about 80% of the warming effect from greenhouse gases, though it has accomplices like methane, nitrous oxide, and various industrial gases (NOAA Climate.gov).

2. It's primarily from fossil fuels, with deforestation, cement production, and other sources playing supporting roles in the climate drama (IPCC AR6).

3. It stays in the atmosphere for centuries, meaning our emissions today will influence Earth's climate for generations to come. It's the ultimate intergenerational IOU (NASA).

4. Solutions exist and are improving rapidly, from renewable energy to electrification to carbon removal, though deploying them at sufficient scale and speed remains challenging (IPCC).

5. The obstacles are primarily political and psychological rather than technological, requiring new policies, economic arrangements, and ways of thinking about our relationship with energy and the environment (WEF).

The story of carbon dioxide is, in many ways, the story of modern human civilization—our incredible technological progress, our global interconnectedness, our creativity, and our sometimes shortsighted pursuit of growth without considering long-term consequences.

By understanding this tiny molecule and its outsized impact, we can better appreciate both the scale of the climate challenge and the pathways to addressing it. Carbon dioxide may be a drama queen, but we're the ones who gave it the spotlight. It's time to redirect our attention to the solutions that can help us write a better ending to this story.

Chapter 4: The Oceans Are Not Okay

They're warming, rising, acidifying, and generally over us.

If Earth had a therapist, the oceans would have a lot to unpack. "I feel taken for granted," they might say from the couch. "Everyone uses me as their personal waste disposal, heat sponge, and all-you-can-fish buffet, but nobody asks how I'm doing. And I'm not doing well. I'm hot, I'm bloated, I'm increasingly acidic, and my circulation isn't what it used to be. Also, there's plastic everywhere."

Covering 71% of Earth's surface and containing 97% of our planet's water, oceans are the true MVPs of our climate system. They absorb about 90% of the excess heat from global warming and roughly 30% of human-caused carbon dioxide emissions (IPCC, 2019). Without the oceans, climate change would be proceeding at a much faster and more obvious pace. It's like having a roommate who's been quietly paying 90% of your shared rent without complaining—until now, when they've reached their limit and are about to have a spectacular meltdown.

So what exactly is happening to our oceans as the climate changes? Grab your metaphorical diving gear—we're going deep on ocean warming, sea level rise, acidification, and changing circulation patterns. Spoiler alert: none of it is good news for beachfront property values.

Ocean Warming: Taking the Heat for Our Carbon Addiction

When we talk about global warming, we typically focus on air temperatures. But here's a crucial fact: about 90% of the heat trapped by greenhouse gases is being absorbed by the oceans (IPCC, 2019). The atmosphere might be getting the headlines, but the oceans are doing the heavy lifting when it comes to dealing with our excess heat production.

Why does so much heat end up in the oceans? Water has an incredible capacity to absorb and store heat—roughly 1,000 times greater than air for the same volume. The oceans are essentially Earth's largest heat sink, slowing atmospheric warming by absorbing vast amounts of thermal energy. It's like having a giant water bottle that absorbs most of the heat from your feverish forehead—great for your forehead in the short term, but eventually that water bottle is going to get uncomfortably hot.

Just how much heat are we talking about? According to the best estimates, the oceans have absorbed heat equivalent to the vast majority of the excess heat in the climate system (IPCC, 2019). That's an enormous amount of thermal energy. If that doesn't qualify as a "heated relationship" with our planet, I don't know what does.

This ocean warming isn't evenly distributed—either geographically or by depth. Surface waters warm faster than the deep ocean, and some regions, like the North Atlantic and Southern Ocean, are absorbing more heat than others. But eventually, this heat makes its way deeper through ocean mixing processes. Recent studies show warming now detectable at depths of 2,000 meters and beyond, meaning even the deep ocean is no longer a refuge from our carbon-fueled fever (IPCC, 2019).

The consequences of ocean warming go far beyond making your beach vacation a few degrees toastier:

1. **Marine heatwaves**: Prolonged periods of abnormally high ocean temperatures are becoming more frequent and intense.

Marine heatwaves have doubled in frequency since 1982 and are increasing in intensity (IPCC, 2019). Imagine a heat wave that lasts for months and kills everything in your neighborhood that can't flee—that's a marine heatwave for coral reefs and other stationary sea life.

2. **Coral bleaching**: When water temperatures get too high, corals expel their symbiotic algae (which provide them with food through photosynthesis), turning white and often dying if the stress continues too long. It's like if you got so hot that you ejected your digestive system—not a sustainable survival strategy.

3. **Shifting marine species**: Many fish and other marine organisms are moving toward the poles or into deeper waters in search of their preferred temperature ranges (NOAA Fisheries, 2024). Great for fishing industries in Alaska, terrible for those in the tropics.

4. **Intensified hurricanes**: Warmer ocean waters provide more energy for tropical storms, potentially increasing their intensity (IPCC, 2021). It's like switching the fuel in your car from regular to premium, except instead of better engine performance, you get more catastrophic wind damage.

5. **Disrupted weather patterns**: Ocean temperatures influence atmospheric circulation and weather systems worldwide. El Niño and La Niña events, which can cause droughts, floods, and temperature anomalies globally, may become more intense or frequent with continued ocean warming.

Sea Level Rise: The Slow-Motion Disaster Already in Progress

If ocean warming is the fever, sea level rise is the swelling that comes with it. Global sea levels are currently rising at about 3.6 millimeters per year—more than twice the rate of the 20th century (IPCC, 2019). That might not sound like much (less than the width of your pinky fingernail each year), but it adds up over time and accelerates as warming continues.

Sea level rise has two primary causes, both linked to global warming:

1. **Thermal expansion**: As water warms, it expands. About one-third of current sea level rise comes from this simple physics principle. It's like how that bottle of olive oil in your pantry mysteriously expands beyond its container on a hot day, except instead of a small oil spill in your cabinet, we're talking about inundating coastlines worldwide.

2. **Melting land ice**: Glaciers and ice sheets on land add water to the oceans when they melt or calve into the sea (IPCC, 2019). This accounts for about two-thirds of current sea level rise. The biggest contributors are mountain glaciers worldwide and the massive ice sheets covering Greenland and Antarctica. To be clear: sea ice melting (like in the Arctic Ocean) doesn't significantly affect sea level since that ice is already floating in the water. It's the ice on land that's the main concern—like a giant ice cube tray that's been accumulating frozen water for thousands of years suddenly defrosting and draining into your bathtub.

How much could sea levels rise? By 2100, the expected range is between 0.3-2.5 meters (1-8 feet), depending on how quickly we reduce emissions and how sensitive ice sheets prove to be (IPCC, 2019). Even the lower end of this range would seriously threaten many coastal areas. The upper end would be catastrophic for many of the world's coastal cities and island nations.

And it doesn't stop in 2100. Once set in motion, ice sheet disintegration can continue for centuries (IPCC, 2019). If all of Greenland's ice melted, sea levels would rise by about 7 meters (23 feet). The complete loss of Antarctica's ice sheets would raise sea levels by about 60 meters (200 feet). That full melting would take thousands of years, but even partial losses this century would fundamentally reshape coastlines worldwide.

The impacts of sea level rise are already being felt:

1. **Coastal flooding**: Many cities are seeing increased "sunny day flooding" during high tides, even without storms (IPCC, 2019). Miami Beach, Norfolk, and other coastal areas are spending billions on adaptation measures like raised roads and enhanced drainage systems. It's like having to renovate your home because the bathtub now randomly overflows every couple of weeks despite nobody using it.

2. **Saltwater intrusion**: Rising seas push saltwater into coastal aquifers, threatening freshwater supplies for millions of people. Imagine going to get a glass of drinking water and finding it's become mysteriously salty—that's the reality facing many coastal communities.

3. **Habitat loss**: Coastal wetlands, beaches, and low-lying islands face inundation, threatening unique ecosystems and biodiversity

(EPA, 2022). Some low-lying island nations like Kiribati and Tuvalu are considering complete relocation of their populations as a last resort.

4. **Property and infrastructure damage**: Trillions of dollars of coastal real estate and critical infrastructure are at risk from rising seas. If you think real estate prices in your area are dropping, wait until the property is literally underwater.

5. **Amplified storm impacts**: Higher sea levels mean storm surges reach further inland, causing more damage (IPCC, 2019). Hurricane Sandy's devastating impact on New York City in 2012 was made worse by the roughly 30 centimeters (1 foot) of sea level rise that had already occurred since 1900. Every inch matters when you're trying to keep water out of subway tunnels and electrical systems.

Ocean Acidification: The Evil Twin of Climate Change

While ocean warming and sea level rise get most of the attention, there's another equally serious consequence of our CO_2 emissions: ocean acidification. About 25-30% of human-caused carbon dioxide emissions dissolve into the ocean, where they form carbonic acid, increasing ocean acidity (NOAA Fisheries, 2024). Since the Industrial Revolution, ocean acidity has increased by about 30%—a rate not seen in at least 65 million years.

To understand ocean acidification, we need a brief chemistry refresher (I promise not to make you balance any equations). When carbon dioxide dissolves in seawater, it undergoes a chemical reaction:

$$CO_2 + H_2O \rightarrow H_2CO_3 \text{ (carbonic acid)} \rightarrow H^+ + HCO_3^-$$

Those H^+ ions increase the acidity of the water (pH is a measure of hydrogen ion concentration). The oceans aren't turning into battery acid —they're still slightly alkaline—but they're becoming less alkaline and more acidic. It's like how your mouth isn't exactly acidic, but it becomes more so after drinking orange juice, leading to that characteristic grimace if you brush your teeth immediately afterward.

This increasing acidity is particularly problematic for marine organisms that build shells or skeletons out of calcium carbonate, including:

1. **Coral reefs**: The foundation of the most diverse marine ecosystems, supporting about 25% of all marine species despite covering less than 1% of the ocean floor. The double whammy of acidification and warming threatens to eliminate most coral reefs by mid-century under high-emission scenarios.

2. **Shellfish**: Oysters, clams, mussels, and other commercially important species face difficulties forming and maintaining their shells in more acidic waters. Some shellfish hatcheries on the U.S. West Coast have already experienced significant production problems linked to acidification (NOAA, 2024).

3. **Plankton**: Many tiny floating organisms form calcium carbonate structures and serve as the base of marine food webs. Their decline could ripple through entire ecosystems. It's like removing the foundation from a building and expecting the upper floors to somehow remain intact.

The process works like this: more acidic water contains fewer carbonate ions, which are essential building blocks for calcium carbonate shells and skeletons. Additionally, existing calcium carbonate structures can begin to dissolve in more acidic conditions, like leaving a seashell in vinegar. It's a bit like trying to build a house while someone is simultaneously removing bricks from your foundation.

Laboratory and field studies show that many marine species experience negative effects when exposed to the acidity levels projected for later this century:

- **Reduced calcification rates**: Many shell-forming organisms build their structures more slowly or with weaker materials in more acidic water.
- **Developmental abnormalities**: The larvae of many marine species show deformities or delayed development.
- **Behavioral changes**: Some fish species exhibit altered behavior, including reduced ability to detect predators or prey.
- **Ecosystem shifts**: As vulnerable species decline, they may be replaced by species better adapted to higher acidity, fundamentally altering marine communities (NOAA, 2024).

Unlike some climate impacts that might be reversed relatively quickly if emissions decline, ocean acidification will persist for thousands of years due to the slow mixing of ocean waters and the even slower geological processes that eventually neutralize the excess acidity. We're essentially conducting a long-term chemistry experiment with our oceans, with marine life as the unwitting test subjects.

Changing Ocean Circulation: Rerouting Earth's Vital Currents

Ocean circulation is like Earth's cardiovascular system—it distributes heat, oxygen, nutrients, and carbon throughout the planet. Climate change is affecting these circulation patterns in complex ways that could have far-reaching consequences for weather patterns and marine ecosystems worldwide.

One of the most important components of global ocean circulation is the Atlantic Meridional Overturning Circulation (AMOC), which includes the Gulf Stream. This system acts like a conveyor belt, carrying warm, salty water northward in the Atlantic's upper layers and returning colder water southward in deeper layers. It's a crucial heat distributor that keeps Western Europe much warmer than it would otherwise be at its latitude (NOAA, 2024).

Climate change affects this circulation through two primary mechanisms:

1. **Warming surface waters**: As the ocean surface warms, the density difference between surface and deep waters increases, potentially weakening the overturning circulation that depends on dense surface waters sinking in the North Atlantic.

2. **Freshwater input**: Melting ice from Greenland and increased precipitation in the North Atlantic add freshwater to the ocean surface, reducing its density and further inhibiting the sinking process that drives the circulation (NCEI, 2024).

Observations indicate the AMOC has already weakened by about 15% since the mid-20th century. Climate models project further weakening of 10-60% by 2100, depending on emissions scenarios. Some models even suggest the possibility of a complete shutdown under extreme warming, though this is considered unlikely this century (Met Office, 2024).

A significant weakening of the AMOC would have major implications:

1. **Cooling in the North Atlantic region**: Despite global warming, parts of Europe could experience regional cooling if the Gulf Stream delivers less tropical heat northward. It would be the ultimate climate irony—some regions actually getting colder because of global warming.

2. **Shifting precipitation patterns**: The position of the Intertropical Convergence Zone (ITCZ), which affects rainfall across the tropics, is influenced by the temperature difference between the Northern and Southern Hemispheres, which in turn is affected by ocean circulation.

3. **Sea level changes**: A weakened AMOC would cause regional sea level rise along North America's east coast and parts of Europe.

4. **Marine ecosystem disruption**: Changes in nutrient delivery and temperature patterns would affect marine productivity and species distributions (Met Office, 2024).

Other important ocean circulation patterns are also changing:

- **Strengthening Western Boundary Currents**: Major currents like the Gulf Stream, Kuroshio, and East Australian Current are intensifying and shifting poleward, affecting regional weather and marine ecosystems.

- **Expanding tropical zones**: The ocean's tropical zones appear to be expanding poleward, with potential impacts on precipitation patterns and marine species ranges.

- **Changing upwelling patterns**: Coastal upwelling systems, which bring nutrient-rich deep water to the surface and support some of the world's most productive fisheries, may intensify in some regions and weaken in others due to changing wind patterns.

The Compounding Crisis: When Ocean Threats Converge

As if each individual ocean impact weren't bad enough, the real disaster comes from their interactions. The combined effects of warming, acidification, sea level rise, circulation changes, deoxygenation, and other stressors like overfishing and pollution create a perfect storm for marine ecosystems:

1. **Coral reef collapse**: The one-two punch of warming (causing bleaching) and acidification (hindering calcification) threatens to eliminate most of the world's coral reefs by mid-century under high-emission scenarios. Since reefs support about 25% of all marine species, their loss would be catastrophic for marine biodiversity and the hundreds of millions of people who depend on them for food, income, and coastal protection.

2. **Compounding coastal threats**: Sea level rise, more intense storms, and the loss of protective ecosystems like coral reefs and mangroves combine to drastically increase coastal vulnerability. It's like removing your home's roof during hurricane season while simultaneously weakening its foundation —multiple failures converging toward catastrophe.

3. **Fisheries decline**: The combined impacts of warming, acidification, deoxygenation, and changing circulation patterns threaten global fisheries that provide the primary source of

animal protein for more than a billion people worldwide (NOAA Fisheries, 2024).

4. **Feedback loops**: Some ocean changes can amplify climate change itself. For example, warming reduces the ocean's ability to absorb carbon dioxide, which would accelerate atmospheric CO_2 increase. Similarly, if warming triggers large releases of methane from undersea deposits, it could significantly accelerate global warming.

5. **Ecosystem tipping points**: Marine ecosystems may be able to withstand a certain level of stress before reaching a tipping point where they rapidly transform into fundamentally different states, potentially with much lower biodiversity and productivity.

The ocean's vast size once made it seem immune to human influence, capable of diluting any pollution or absorbing any impact. We now know better. Despite covering more than 70% of Earth's surface and having an average depth of about 3,700 meters (12,100 feet), the world's oceans are showing unmistakable signs of strain from human activities. It's like discovering that your seemingly indestructible older sibling is actually vulnerable after all—both a sobering realization and a call to be more careful.

Blue Hope: Ocean Solutions for the Climate Crisis

Despite these serious threats, there are reasons for optimism. The oceans offer numerous opportunities for climate solutions:

1. Marine Protected Areas: Underwater Conservation

Establishing and enforcing marine protected areas (MPAs) can enhance ecosystem resilience to climate impacts while protecting biodiversity and carbon stores in coastal habitats like mangroves, seagrass beds, and salt marshes. These "blue carbon" ecosystems sequester carbon at rates up to 100 times greater than forests on land, making their protection a climate double win (The Blue Carbon Initiative, 2024).

Currently, only about 7% of ocean areas have some form of protection, with less than 3% fully protected from extractive and destructive activities. Increasing this to the 30% target advocated by many scientists would significantly boost ocean health and resilience (NOAA Sanctuaries, 2024).

2. Sustainable Fisheries: Food Security in a Changing Ocean

Reforming fisheries management to account for climate impacts can ensure continued food security while reducing additional stress on marine ecosystems. This includes:

- Adjusting catch limits to account for changing productivity
- Protecting critical habitat for spawning and juvenile development
- Shifting to more selective fishing gear that reduces bycatch
- Developing international agreements for managing shifting fish stocks

These approaches can maintain the ocean's role as a critical food source for billions while allowing ecosystems to better withstand climate pressures (NOAA Fisheries, 2024).

3. Ocean-Based Renewable Energy: Power from the Blue

The oceans offer vast renewable energy potential through:

- **Offshore wind**: Technology has advanced dramatically, with floating turbines now enabling deployment in deeper waters with stronger, more consistent winds.
- **Wave and tidal energy**: Though less developed than wind, these technologies are advancing and could provide reliable, predictable renewable power.
- **Ocean thermal energy conversion (OTEC)**: Using the temperature difference between warm surface waters and cold deep waters to generate electricity—particularly promising for tropical island nations.

These technologies can help decarbonize our energy systems while creating jobs in coastal communities.

4. Coastal Adaptation: Living With Rising Seas

Innovative approaches to coastal adaptation include:

- **Living shorelines**: Using natural elements like oyster reefs, salt marshes, and mangroves to reduce erosion and flood risks while enhancing habitat.
- **Floating architecture**: Developing buildings and infrastructure designed to rise and fall with water levels.
- **Managed retreat**: Strategically relocating development away from high-risk areas while creating new ecological and recreational spaces in the process.

These approaches recognize that some sea level rise is now inevitable and focus on adapting intelligently rather than fighting a losing battle against the rising tides.

5. Ocean Alkalinization: Countering Acidification

Research is exploring ways to counter ocean acidification locally through:

- **Adding alkaline materials**: Introducing substances like limestone or olivine to increase water alkalinity and buffer against acidification.
- **Seaweed cultivation**: Growing macroalgae that absorb CO_2 during photosynthesis and potentially reduce local acidity.
- **Electrochemical approaches**: Using low-voltage current to trigger chemical reactions that increase water alkalinity.

While still largely experimental, these approaches could potentially protect particularly vulnerable or valuable marine areas like coral reefs and shellfish beds.

6. Reducing Land-Based Pollution: Cutting Other Stressors

Reducing non-climate stressors on ocean ecosystems increases their resilience to climate impacts:

- **Nutrient pollution reduction**: Addressing agricultural runoff and sewage discharges that cause harmful algal blooms and dead zones.
- **Plastic pollution prevention**: Implementing comprehensive strategies to stop plastic from entering marine environments.
- **Chemical contaminant control**: Strengthening regulations on persistent pollutants that accumulate in marine food webs.

These measures give marine ecosystems a fighting chance against climate impacts by reducing the total stress burden they face (EPA, 2022).

The Ocean-Climate Connection: Why Ocean Health Is Climate Health

The oceans and climate are inextricably linked in ways that make ocean conservation a climate imperative. Healthy oceans regulate climate, while degraded oceans amplify climate risks. This connection works in several ways:

1. **Carbon sequestration**: The ocean's biological pump transfers carbon from the surface to the deep ocean, where it can remain sequestered for centuries to millennia. Supporting this natural carbon sink through ocean conservation helps mitigate climate change.

2. **Heat absorption**: By absorbing and distributing excess heat, the oceans moderate climate extremes. Maintaining ocean circulation patterns through emissions reductions helps preserve this crucial service.

3. **Weather regulation**: Ocean temperatures and circulation patterns influence weather systems worldwide, from monsoons to hurricanes to seasonal precipitation. Climate stability depends on ocean stability.

4. **Food and livelihood security**: As climate change threatens terrestrial agriculture in many regions, sustainable ocean food production becomes increasingly important for global food security (NOAA Fisheries, 2024).

These connections mean that protecting the oceans isn't just about saving marine life—it's about safeguarding human well-being and stabilizing our climate system. It's a classic "what goes around comes around" situation: the better we treat the oceans, the better they can help us weather (literally) the climate challenges ahead.

Conclusion: Why We Need to Make Waves About Ocean Health

The oceans have been quietly buffering us from the worst impacts of climate change, absorbing heat and carbon dioxide that would otherwise be dramatically warming our atmosphere. But this marine heroism comes at a cost—increasingly acidic waters, rising sea levels, disrupted circulation patterns, and threatened ecosystems.

The good news is that oceans are remarkably resilient. Given a chance to recover, marine ecosystems can bounce back and continue providing the essential services that human civilization depends on. But that resilience isn't infinite, and we're testing its limits with the combined pressures of climate change, overfishing, pollution, and habitat destruction.

Ocean conservation and climate action aren't separate challenges—they're two sides of the same coin. By reducing greenhouse gas emissions, we slow ocean warming, acidification, and sea level rise. By protecting and restoring marine ecosystems, we enhance natural carbon sinks and build resilience to unavoidable climate impacts.

The fate of the oceans is ultimately our fate as well. We evolved on a planet whose surface is mostly ocean, where every breath we take contains oxygen produced in part by marine phytoplankton, and where the water cycle that provides our drinking water depends on healthy ocean systems. We're not so much land animals who occasionally visit the ocean as we are ocean animals who occasionally visit the land.

So next time you stand at the shore watching waves crash onto the beach, remember: those seemingly eternal waters are changing because of us, and their future depends on the choices we make today. The ocean may be vast, but it's not infinite. It's time we treated it with the care and respect it deserves—not just for the sake of the incredible life it contains, but for our own survival on this blue planet.

Chapter 5: How We Know What We Know: Satellites, Trees, and Nerds

Also, those "hockey stick" graphs aren't about sports.

Let's address the elephant in the room: how do we actually *know* all this climate stuff? How can scientists claim to understand what Earth's climate was like thousands of years ago, let alone predict what might happen decades from now? Are they wizards? Time travelers? Really good guessers?

The answer is both more mundane and more fascinating than magic: it's methodical detective work using a diverse array of scientific tools, techniques, and evidence. Climate science is like a massive crime scene investigation where the victim is our stable climate, and the suspects include a lineup of greenhouse gases with carbon dioxide as the primary culprit. The scientific community has assembled an overwhelming case file against these climate-altering substances, and in this chapter, we'll examine the evidence.

From tree rings that record centuries of growing conditions to ice cores that capture ancient air bubbles, from satellites that measure Earth's energy balance to sophisticated computer models that simulate the entire climate system, climate scientists have developed an impressive toolkit for understanding our planet. So put on your deerstalker cap and grab your magnifying glass—we're about to explore how scientists know what they know about climate change.

Time Machines in Ice: The Frozen Climate Archives

One of the most remarkable tools in climate science is the humble ice core—a cylinder of ice drilled from the ice sheets of Greenland or Antarctica, or from high-mountain glaciers. These frozen records contain layers like the rings of a tree, with each layer representing a season of snowfall that was gradually compressed into ice.

The oldest continuous ice core records stretch back about 800,000 years (from Antarctica), giving us an unprecedented window into Earth's climate history (NSIDC, "What do ice cores reveal about the past?"). That's like having a detailed diary of Earth's climate that started when early humans were just beginning to use fire regularly. It's a planetary autobiography written in frozen water.

What makes ice cores so valuable is what gets trapped inside them. As snow falls and eventually compresses into ice, it captures:

1. **Air bubbles**: Small pockets of atmosphere get trapped in the ice, preserving the actual composition of the air from when that layer formed. Scientists can directly measure the concentration of greenhouse gases like CO_2 and methane in these ancient air samples. It's like having bottled atmosphere from thousands of years ago—the ultimate vintage air collection.

2. **Isotopes of oxygen and hydrogen**: The ratio of heavy to light isotopes in the water molecules of the ice provides a remarkably accurate thermometer of past temperatures. Warmer conditions lead to more heavy isotopes in precipitation, while colder conditions favor lighter isotopes.

This relationship allows scientists to reconstruct past temperatures with surprising precision.

3. **Dust and other particles**: Volcanic ash, desert dust, sea salt, pollen, and even pollution from human activities get deposited on the ice and preserved in the layers. These serve as markers of past environmental conditions and events.

The ice core record tells a clear and consistent story: current CO_2 levels (above 410 ppm) are far higher than anything seen in the past 800,000 years, when concentrations oscillated between about 180 ppm during ice ages and 280 ppm during warmer interglacial periods (Global Climate Change Impacts, "800,000 Year record of CO2 Concentration"). Our current situation is literally off the charts of the ice core record.

Moreover, the rate of CO_2 increase is unprecedented—rising by about 100 ppm in just the past 60 years, compared to typical natural changes of about 100 ppm over 10,000 years or more during glacial-interglacial transitions (NOAA Climate.gov, "Climate Change: Atmospheric Carbon Dioxide"). It's like comparing someone gradually gaining weight over decades versus someone binge-eating their way to the same weight gain in a single month.

Tree Detectives: Reading Nature's Growth Rings

While ice cores provide our longest continuous climate records, they're limited to regions where ice persists year-round. For the rest of the world, particularly the temperate and tropical regions where most people live, we need other proxies. Enter dendroclimatology: the science of reconstructing past climate from tree rings.

Trees are nature's record-keepers, adding a new growth ring each year. The width, density, and chemical composition of these rings reflect the growing conditions the tree experienced—including temperature, precipitation, and even the chemical composition of the atmosphere (NOAA Climate.gov, "How tree rings tell time and climate history"). Some long-lived species like bristlecone pines can live for thousands of years, while archaeological remains and preserved logs can extend tree ring records even further back in time.

What makes tree rings especially valuable is their precise annual resolution and widespread geographical distribution. Unlike ice cores, which are limited to polar and high-mountain regions, trees grow on every continent except Antarctica, allowing scientists to reconstruct regional climate patterns across much of the globe.

Tree rings have been particularly important in constructing temperature records for the past 2,000 years, including the famous "hockey stick" graph that shows recent warming is unprecedented in both rate and magnitude compared to natural variations of the past millennium (IPCC, "Climate Change 2021: The Physical Science Basis"). The "hockey stick" nickname comes from the shape of the graph—a relatively flat "handle" representing temperature variations over the past 1,000 years, followed by a sharp upward "blade" in the 20th century as human-caused warming kicked in.

This graph became controversial not because of scientific flaws (it has been repeatedly confirmed by subsequent studies using various methods), but because of its clear implication: recent warming is not part of a natural cycle but a dramatic departure caused by human activities. Some people really don't like that conclusion, much like how certain members of my family don't like it when I bring up their questionable financial decisions at Thanksgiving dinner.

Beyond temperature, tree rings also record drought conditions, wildfire history, and even the effects of volcanic eruptions. For example, unusually narrow rings often appear following major volcanic eruptions, which can temporarily cool the climate by injecting aerosols into the stratosphere that reflect sunlight. It's like finding evidence of your teenager's party in the recycling bin—the empty bottles tell a story even when no one's talking.

Sediment Sleuths: Mud with a Message

When trees and ice aren't available, scientists turn to mud—specifically, the layered sediments that accumulate at the bottoms of lakes and oceans. Like ice cores and tree rings, many sediments form distinct annual layers (called varves) that can be counted to establish a chronology. Within these layers, scientists find a treasure trove of climate indicators:

1. **Microfossils**: The remains of tiny organisms like foraminifera (marine single-celled creatures with shells), diatoms (algae with glass-like cell walls), and pollen grains preserve information about temperature, ocean chemistry, and vegetation patterns of the past.

2. **Chemical signatures**: The ratios of various elements and isotopes in sediments and the shells of organisms reflect environmental conditions when they formed, providing proxies for temperature, precipitation, ocean circulation, and more.

3. **Physical properties**: The size, sorting, and composition of sediment particles reveal information about past storm intensity, river flows, glacier extent, and sea level.

Ocean sediments have been particularly valuable for reconstructing Earth's longer-term climate history, extending back tens of millions of years. The consistency of ocean conditions allows for more continuous records than terrestrial environments, which are more prone to erosion and disturbance.

One of the most important findings from ocean sediment records is the close correlation between CO_2 levels and global temperature throughout Earth's history. Periods of high CO_2 consistently correspond to warmer climates, while periods of low CO_2 correspond to cooler climates (Carbon Brief, "Explainer: How the rise and fall of CO2 levels influenced the ice ages"). This relationship holds across timescales from decades to millions of years, providing powerful evidence for CO_2's role as a major driver of Earth's climate.

Sediments also record rapid climate shifts of the past, such as the Paleocene-Eocene Thermal Maximum (PETM) about 56 million years ago, when a massive release of carbon into the atmosphere caused global warming of 5-8°C over just a few thousand years. This ancient climate disruption serves as a sobering analog for current human-caused climate change, though we're adding carbon to the atmosphere at least 10 times faster than during the PETM (NASA, "The Last Time the Globe Warmed"). It's like comparing a fender bender to a high-speed collision—the basic mechanism is the same, but the consequences are far more severe at higher speeds.

Coral Chronicles: Reefs as Climate Recorders

Like trees on land, corals in the ocean create annual growth bands that preserve information about past climate conditions. By analyzing the chemistry and structure of these bands, scientists can reconstruct sea surface temperatures, rainfall patterns, and even the strength of ocean currents going back centuries to millennia.

The ratio of oxygen isotopes in coral skeletons varies with water temperature and salinity, providing a record of past sea surface conditions. Similarly, the amounts of certain trace elements like strontium and calcium vary with temperature, offering independent confirmation of temperature reconstructions (NOAA, "Coral Paleoclimate Research").

Corals have proven particularly valuable for reconstructing climate patterns in the tropics, where other high-resolution proxies may be limited. For example, coral records from the tropical Pacific have helped scientists understand the history of El Niño and La Niña events, which have far-reaching impacts on global weather patterns.

Unfortunately, the same human activities that are disrupting Earth's climate are also threatening the corals that help us understand it. Ocean warming and acidification are causing widespread coral bleaching and mortality, potentially limiting our ability to extend coral-based climate reconstructions into the future (NOAA, "What is coral bleaching?"). It's like burning the history books while we're still trying to learn from them.

The Modern Arsenal: Satellites, Stations, and Sensors

While paleoclimate records are essential for understanding Earth's climate history, modern instrumental measurements provide unprecedented detail about current climate conditions and recent changes. The modern climate monitoring system includes:

1. **Weather stations**: Thousands of surface stations worldwide measure temperature, precipitation, wind, humidity, and air pressure. Some of these records extend back to the 1800s, providing our longest direct observations of climate variables (NOAA, "Climate data monitoring").

2. **Weather balloons and aircraft**: Regular launches of instrumented balloons and specialized research aircraft measure conditions throughout the atmosphere, not just at the surface.

3. **Ocean instruments**: Networks of buoys, autonomous underwater vehicles, and ship-based measurements track ocean temperatures, salinity, currents, and chemistry. The Argo array of nearly 4,000 floating instruments provides comprehensive monitoring of the upper ocean worldwide.

4. **Satellites**: Since the late 1970s, Earth-observing satellites have revolutionized climate monitoring by providing truly

global coverage (NASA, "Climate Change"). They measure surface and atmospheric temperatures, sea levels, ice extent, atmospheric composition, vegetation patterns, and many other variables critical to understanding climate.

This integrated observing system provides a comprehensive, real-time view of Earth's climate that would have been unimaginable to previous generations. It's like going from checking your bank account once a month with a paper statement to having a real-time finance app on your phone that tracks every penny—the increased detail reveals patterns that would otherwise remain hidden.

Of all these modern tools, satellites have perhaps been most transformative. They provide truly global coverage, including remote regions like the open ocean, deserts, and polar regions where surface measurements are sparse (NOAA, "How are satellites used to observe the ocean?"). They observe Earth consistently using the same instruments, avoiding the complications that arise when comparing measurements from different ground-based stations with varying equipment and surroundings.

Satellite capabilities particularly relevant to climate monitoring include:

- **Measuring Earth's energy balance**: Satellites like NASA's CERES (Clouds and the Earth's Radiant Energy System) directly measure how much solar energy Earth absorbs and how much it radiates back to space—the fundamental driver of climate change.

- **Tracking ice changes**: Satellites monitor the extent, thickness, and mass of ice sheets, sea ice, and glaciers worldwide with unprecedented precision, documenting their rapid changes in response to warming.

- **Measuring sea level**: Satellite altimeters measure global sea level with millimeter-scale precision, revealing not just the global average rise but also regional variations due to ocean currents, land movements, and gravitational effects.

- **Observing the atmosphere**: Satellites measure the concentration and distribution of greenhouse gases, aerosols, water vapor, and other atmospheric constituents that influence climate.

Contrary to claims by some climate skeptics, satellite temperature measurements confirm the warming trend observed in surface-based records. While there were discrepancies in early satellite temperature analyses due to issues like orbital drift and instrument calibration, these have been resolved, and all major datasets now show consistent warming trends, regardless of whether they're based on satellites, weather balloons, or surface stations (NASA, "Satellite Measurements of Warming in the Troposphere"). It's like getting a second and third medical opinion that confirm the initial diagnosis—at some point, you should probably start treatment instead of doctor-shopping for someone who'll tell you everything is fine.

Climate Models: Digital Earths in Silicon

While observations tell us what has happened and is happening to Earth's climate, computer models help us understand why these changes occur and what might happen in the future under different scenarios. Climate models are essentially digital representations of Earth's climate system, based on fundamental physical, chemical, and biological principles expressed as mathematical equations.

Modern climate models, also called Earth System Models, simulate:

- The atmosphere's circulation, temperature, humidity, and chemical composition
- Ocean currents, temperature, salinity, and chemistry
- Sea ice formation, movement, and melting
- Land surface processes, including vegetation, soils, and hydrology
- The carbon cycle, including exchange of carbon between the atmosphere, ocean, land, and biosphere
- Human influences, such as greenhouse gas emissions, land use changes, and aerosol pollution

These models divide Earth into a three-dimensional grid and calculate how energy, moisture, momentum, and various substances move between grid cells according to the laws of physics. The most advanced models use millions of grid cells and represent hundreds of interacting processes, requiring supercomputers to run.

Climate models are routinely tested against past climate changes to ensure they realistically simulate Earth's climate dynamics (NOAA,

"Climate Models"). For example, models are evaluated on their ability to reproduce:

- Seasonal temperature cycles
- The effects of major volcanic eruptions
- El Niño and La Niña patterns
- The climate response to changes in Earth's orbit that drive ice age cycles
- Historical climate changes over the 20th century

This rigorous testing gives scientists confidence that models can provide useful insights about future climate under various emission scenarios. It's like how we trust weather forecasting models enough to plan our weekends, but with much longer time horizons and different types of predictions.

It's important to understand what climate models are (and aren't) used for. They don't predict exactly what the temperature will be in Chicago on July 4, 2050—that's weather forecasting, not climate modeling. Rather, they project how average conditions and the frequency of extreme events will change under different greenhouse gas concentration pathways.

Climate models aren't perfect—they still struggle with representing some processes like clouds and precipitation at local scales. But they've proven remarkably skilled at projecting large-scale temperature patterns and trends. In fact, climate models from the 1970s and 1980s successfully predicted the warming we've experienced over the past several decades, despite using much simpler computers and having less data than today's models (NASA, "Climate Models").

Fingerprinting Climate Change: How We Know Humans Are Responsible

Perhaps the most crucial question in climate science isn't whether Earth is warming (it is) or whether CO_2 affects climate (it does), but whether humans are the primary cause of recent changes. This is where attribution science comes in—the field dedicated to identifying the fingerprints of human influence on the climate system.

Scientists use several approaches to distinguish human-caused climate change from natural variability:

1. **Pattern analysis**: Human-caused warming has a distinctive spatial and temporal pattern that differs from natural variability. For example, human greenhouse gas emissions warm the troposphere (lower atmosphere) while cooling the stratosphere (upper atmosphere)—a signature that natural factors like solar variability or volcanoes don't produce (IPCC, "Climate Change 2021: The Physical Science Basis").

2. **Formal detection and attribution studies**: These use statistical techniques to determine whether observed changes exceed what would be expected from natural variability alone, and then assess the contribution of various factors (greenhouse gases, solar changes, volcanoes, etc.) to the detected changes.

3. **Process understanding**: By quantifying the physical mechanisms through which various factors affect climate, scientists can determine their relative importance in driving observed changes.

The evidence from these approaches overwhelmingly indicates that humans are the dominant cause of warming since the mid-20th century (UN, "Global warming 'unequivocally' human driven: IPCC"). Natural factors like solar variability and volcanoes have been thoroughly investigated and found insufficient to explain observed warming trends. In fact, natural factors alone would likely have produced slight cooling over recent decades, meaning human activities have driven more than 100% of the observed warming (counteracting a small natural cooling tendency).

Specific human fingerprints on the climate system include:

- **The vertical pattern of atmospheric warming**: Greenhouse gases trap heat in the lower atmosphere, warming the troposphere while cooling the stratosphere— exactly the pattern observed by weather balloons and satellites.

- **The geographic pattern of warming**: Models predict stronger warming over land than oceans, and greater warming in the Arctic than the tropics—patterns clearly seen in observations.

- **The isotopic composition of atmospheric CO_2**: Fossil fuel carbon has a distinctive isotopic signature (less carbon-13 and no carbon-14), and measurements show that atmospheric CO_2 bears this signature, directly linking rising CO_2 levels to fossil fuel combustion (NOAA, "Carbon Isotopes").

- **Cooling of the upper atmosphere**: Enhanced greenhouse warming traps heat in the lower atmosphere, reducing energy flow to the upper atmosphere and causing it to cool—a pattern observed in satellite measurements.

Taken together, these multiple lines of evidence form an ironclad case for human responsibility. It's like a detective novel where all the clues—fingerprints, DNA, eyewitness accounts, security camera footage, and a signed confession—point to the same culprit.

From Global to Local: Downscaling Climate Knowledge

While global climate models provide crucial insights about large-scale changes, many climate impacts and adaptation decisions happen at local to regional scales. Communities need to know how their specific area might be affected by climate change, not just how the global average will shift.

To bridge this gap, scientists use "downscaling" techniques to translate global model projections to finer spatial scales. These approaches include:

1. **Statistical downscaling**: Using observed relationships between large-scale climate patterns and local conditions to infer how local climate will respond to global changes. For example, knowing how a region's rainfall historically responds

to El Niño conditions helps predict its response to future warming.

2. **Regional climate modeling**: Embedding higher-resolution models within global models to capture local details like the influence of mountains, coastlines, and land use patterns on temperature and precipitation.

3. **Hybrid approaches**: Combining statistical and modeling techniques to leverage the strengths of each method.

These downscaling methods help communities understand specific risks like changes in growing seasons, flood probabilities, heat wave frequency, or water availability. They've become essential tools for climate adaptation planning at local scales (EPA, "Downscaling Climate Information").

However, downscaling has limitations. Increasing spatial resolution doesn't necessarily improve accuracy if the underlying global model has biases, and some local climate responses may depend on processes that aren't well-represented even in high-resolution models. It's a bit like digital zoom on a camera—you can make the image bigger, but you can't create detail that wasn't captured in the original picture.

Despite these limitations, downscaled climate information has proven invaluable for adaptation planning. It helps bridge the gap between global climate science and local decision-making, making abstract projections concrete and relevant to communities. When a city planner can see how flood risks in their specific watershed might change, or a farmer can understand how growing seasons in their county might shift, climate adaptation moves from theoretical to practical.

When Data Disagree: Handling Uncertainty in Climate Science

No scientific field achieves perfect certainty, and climate science is no exception. Climate data contain measurement uncertainties, models have limitations, and projections of future change depend on unpredictable human choices about emissions. How do climate scientists handle these uncertainties, and what do they mean for our understanding of climate change?

First, it's important to distinguish different types of uncertainty:

1. **Measurement uncertainty**: All observations have some margin of error. Modern temperature measurements might be accurate to within 0.1°C, while reconstructions from tree rings or ice cores typically have larger uncertainties.

2. **Natural variability**: Earth's climate fluctuates naturally from year to year and decade to decade due to processes like El Niño, volcanic eruptions, and solar cycles. This "noise" can temporarily mask or amplify the "signal" of human-caused warming.

3. **Model uncertainty**: Climate models differ in how they represent certain processes, leading to a range of projections even under identical emissions scenarios.

4. **Scenario uncertainty**: We don't know exactly how human emissions will evolve, dependent as they are on population growth, economic development, technological change, and policy decisions.

Climate scientists address these uncertainties through several approaches:

- **Ensemble modeling**: Rather than relying on a single model run, scientists use multiple models and multiple runs of each model to generate a range of projections that better characterize uncertainty (NOAA, "Climate Model Ensembles").

- **Probabilistic statements**: Instead of predicting a single value for future warming, scientists provide ranges and likelihood estimates, such as "Global temperatures are likely (>66% probability) to increase by 2.5-4°C under a high-emissions scenario."

- **Transparent communication**: The Intergovernmental Panel on Climate Change (IPCC) and other scientific bodies explicitly discuss uncertainties and use standardized language to describe confidence levels and probability ranges.

- **Focusing on robust findings**: Some conclusions emerge consistently despite uncertainties, such as "continued emissions will lead to continued warming" and "the risks of serious impacts increase with greater warming."

Importantly, uncertainty cuts both ways—things could turn out better than expected, but they could also turn out worse. Some climate impacts have occurred faster than projected in earlier assessments, including Arctic sea ice loss, ice sheet melting, and sea level rise. Uncertainty is not a reason for complacency but rather a reason for risk management approaches that consider the full range of possible outcomes, including low-probability but high-consequence scenarios.

As science has progressed, some uncertainties have narrowed. For example, the estimated range for "climate sensitivity" (how much warming results from a doubling of CO_2) has remained stubbornly wide for decades, but recent research combining multiple lines of evidence has begun to constrain it more tightly. Similarly, improved understanding of ice sheet dynamics is refining projections of future sea level rise.

Overall, while uncertainties remain in climate science (as in all sciences), they primarily affect the precise magnitude and timing of impacts rather than the fundamental conclusion that human activities are warming the planet in ways that pose significant risks (NASA, "Responding to Climate Change"). It's like knowing a hurricane is headed for your city—you might not know exactly when it will make landfall or precisely how strong it will be, but that uncertainty doesn't mean you should ignore evacuation warnings.

The Scientific Consensus: Strength in Evidence and Agreement

One of the most misunderstood aspects of climate science is the nature and extent of the scientific consensus. Contrary to common misconceptions, scientific consensus isn't determined by vote, opinion polls, or appeals to authority—it emerges from the convergence of evidence and the testing of competing hypotheses.

On climate change, this evidence-based consensus is overwhelming. Multiple studies have found that about 97-99% of climate scientists and 97-98% of climate-relevant research papers endorse the conclusion that human activities are the primary driver of recent warming. This level of agreement is comparable to the scientific consensus on evolution, plate tectonics, or the link between smoking and cancer (NASA, "Scientific Consensus: Earth's Climate is Warming").

The consensus is strongest on the foundational elements of climate science:

- Earth's climate is warming (essentially 100% agreement)
- Human activities, particularly greenhouse gas emissions, are the primary cause of recent warming (97-99% agreement)
- Continued emissions will lead to continued warming and increasing risks (very high agreement)
- Reducing emissions will reduce future warming and associated risks (very high agreement)

Legitimate scientific debates exist around more detailed questions, such as:

- Exactly how sensitive is the climate system to greenhouse gases?
- How will specific regional precipitation patterns change?
- How quickly will ice sheets respond to warming?
- What role will clouds play in amplifying or dampening warming?

These ongoing research questions reflect the normal process of science refining its understanding, not fundamental uncertainty about the reality or human causation of climate change.

The strength of the scientific consensus lies not just in the percentage of agreeing scientists but in the multiple, independent lines of evidence that lead to the same conclusion. When ice cores, tree rings, satellite measurements, ocean heat content, species migrations, and sophisticated model simulations all point to human-caused warming, the conclusion becomes virtually inescapable.

Critics sometimes mischaracterize scientific consensus as groupthink or dogma, but this fundamentally misunderstands how science works. Scientific consensus emerges despite powerful incentives for scientists to disprove existing theories—after all, disproving a major theory is how scientists make their names and advance their careers. If solid evidence existed to overturn our understanding of human-caused climate change, ambitious scientists would be racing to publish it, not suppressing it.

From Science to Action: Bridging Understanding and Response

Understanding how we know what we know about climate change is more than an intellectual exercise—it's essential for informed decision-making. The methods, evidence, and conclusions of climate science provide the foundation for climate policy, business planning, and individual choices.

The climate science we've explored in this chapter tells us several things with high confidence:

1. **The climate is changing**, primarily due to human greenhouse gas emissions.
2. **The changes are accelerating** as emissions continue to rise.
3. **Many impacts are already occurring**, from rising seas to more intense heat waves.
4. **Future risks depend largely on our choices** about emissions.
5. **Both mitigation (reducing emissions) and adaptation (preparing for changes) are necessary** (EPA, "Climate Change Science").

This scientific understanding doesn't tell us exactly what policies to adopt—those decisions involve values, priorities, and trade-offs that science alone cannot resolve. But science does clarify the consequences of different choices and the urgency of action.

Perhaps the most important thing climate science tells us is that delay is costly. The longer we wait to reduce emissions, the more warming becomes locked in, the more severe the impacts become, and the more limited our future options are. It's like compound interest working against us—the climate debt we're accumulating becomes increasingly difficult to pay off the longer we delay.

Scientists have also clarified the scale of the challenge. To have a good chance of limiting warming to 1.5°C above pre-industrial levels —the aspirational goal of the Paris Agreement—global CO_2 emissions need to fall by about 45% from 2010 levels by 2030 and reach net zero around 2050. For a 2°C limit, emissions need to drop about 25% by 2030 and reach net zero around 2070. Either pathway requires rapid, far-reaching changes across all sectors of the global economy (IPCC, "Climate Change 2021: The Physical Science Basis").

Challenging the Scientists: Playing Ping-Pong with Climate Skeptics

Despite the robust scientific consensus on climate change, a small but vocal chorus of skeptics continues to challenge the findings we've discussed. Let's examine some common skeptical arguments and how they hold up against the scientific evidence:

Skeptic Claim #1: "The climate has always changed naturally, so current warming must be natural too."

The Science Says: This is like arguing that because people die of natural causes, murder doesn't exist. The fact that natural climate changes occurred in the past doesn't mean humans can't cause climate change now. Scientists have thoroughly investigated natural factors like solar variations and volcanic activity and found they cannot explain recent warming. Meanwhile, human greenhouse gas emissions provide a quantitatively accurate explanation of observed changes (NASA, "Climate Change: How Do We Know?").

Skeptic Claim #2: "There's no consensus among scientists about climate change."

The Science Says: Multiple independent studies have found that about 97-99% of climate scientists and climate-relevant research papers endorse the conclusion that human activities are the primary cause of recent warming. The consensus is strongest among scientists with the most expertise in climate science. The few contrarian papers that exist have typically been found to contain methodological errors or flawed assumptions (NASA, "Scientific Consensus: Earth's Climate is Warming").

Skeptic Claim #3: "Climate models are unreliable and can't be trusted."

The Science Says: Climate models have successfully predicted many observed changes, including global warming, Arctic amplification, stratospheric cooling, and increased water vapor. While they're not perfect, especially at regional scales, their track record for large-scale climate projections is actually quite good. Models developed in the 1970s and 1980s successfully predicted the warming we've experienced over the past several decades, despite using much simpler computers and having less data than today's models (NASA, "Climate Models").

Skeptic Claim #4: "CO_2 is a trace gas and can't possibly have a big effect on climate."

The Science Says: This argument confuses abundance with impact. Many substances have powerful effects in small concentrations—just ask anyone who's had food poisoning from a tiny amount of bacteria or gotten drunk from a blood alcohol concentration of less than 0.1%. The heat-trapping properties of CO_2 have been measured precisely in laboratories and observed in action in the atmosphere. The physics is clear and uncontroversial among scientists (NASA, "The Atmosphere: Getting a Handle on Carbon Dioxide").

Skeptic Claim #5: "Scientists are just following the money and saying what gets them funding."

The Science Says: This conspiracy theory fails basic logical scrutiny. If a scientist could legitimately disprove human-caused climate change, they would become instantly famous and receive massive attention and funding opportunities. There would be a Nobel Prize waiting for anyone who could overturn our current understanding with solid evidence. The incentives in science favor overturning established theories, not reinforcing them (Skeptical Science, "What does the full body of evidence tell us about global warming?").

Most importantly, climate skepticism often focuses on small uncertainties or cherrypicked data points while ignoring the overwhelming body of evidence. It's like fixating on a single pixel in a high-definition image while refusing to acknowledge the clear picture that emerges when you look at the whole screen. Scientific understanding comes from integrating multiple lines of evidence, not from isolated factoids or anomalies.

From Understanding to Action: The Path Forward

The science we've explored in this chapter gives us a clear picture of Earth's changing climate. We know our planet is warming primarily due to human greenhouse gas emissions. We know this from multiple, independent lines of evidence—ice cores, tree rings, satellites, weather stations, ocean measurements, and more. We understand the underlying physics and chemistry well enough to predict future changes under different emission scenarios.

This scientific understanding provides the foundation for informed action. It tells us that reducing greenhouse gas emissions can limit future warming and associated risks. It shows that many climate solutions—from renewable energy to reforestation—have strong scientific support. And it clarifies that while uncertainties remain about the precise timing and magnitude of some impacts, the fundamental reality of human-caused climate change is not in serious scientific dispute.

Science alone doesn't tell us exactly what policies to adopt or how to balance competing priorities in responding to climate change. Those decisions involve values, ethics, economics, and politics beyond the scope of scientific analysis. But science does give us the factual basis for making informed choices. It's like how medical science can tell you what smoking does to your lungs but can't decide for you whether to quit—that choice ultimately depends on how you weigh the health risks against whatever benefits you perceive from smoking.

Perhaps the most important message from climate science is one of agency and possibility. While the evidence for human-caused warming is overwhelming, the science also shows that our future climate largely depends on choices we make now and in the coming decades (IPCC, "Climate Change 2021: The Physical Science Basis"). Every ton of emissions avoided matters. Every tenth of a degree of warming prevented reduces risks. The scientific methods and evidence we've explored don't just tell us about climate problems— they help guide us toward climate solutions.

In the next chapter, we'll examine one of the most visible manifestations of our changing climate: the increasing frequency and intensity of extreme weather events. From heat waves and floods to wildfires and hurricanes, we'll explore how climate change is loading the weather dice and what that means for communities worldwide. Spoiler alert: your homeowner's insurance rates are probably going up.

Chapter 6: Heat Waves, Floods & Fire: Oh My!

When extreme weather gets an unwanted performance enhancement

Remember that scene in practically every sports movie where the underdog athlete suddenly starts taking mysterious supplements, grows improbably muscular, and begins demolishing their competition? Climate change is essentially doing the same thing to our weather—except nobody's cheering, there's no inspirational montage, and the "supplements" are greenhouse gases that we've been enthusiastically pumping into the atmosphere since the Industrial Revolution.

Extreme weather events have always been a part of Earth's climate system, like that one dramatic friend who occasionally causes a scene at dinner parties. But climate change is transforming these occasional disruptions into something more akin to living with an entire household of reality TV contestants—dramatic incidents aren't just more common, they're more intense, longer-lasting, and increasingly bizarre.

In this chapter, we'll explore how climate change is supercharging different types of extreme weather, from heat waves that make sidewalks hot enough to fry eggs (please don't actually try this—salmonella doesn't care about your social media clout) to rain events that dump a month's worth of precipitation in a day. It's like Earth's weather got a subscription to a very questionable gym and has been aggressively bulking up ever since.

Heat Waves: When "Hot Girl Summer" Gets Too Literal

Let's start with heat waves—periods of abnormally hot weather that last for days or weeks. These are the most direct and obvious consequence of global warming. As the average temperature increases, the bell curve of temperature distribution shifts toward the warmer end, making previously rare extreme heat much more common. It's like if the "medium" setting on your shower suddenly became what "scalding" used to be, but your roommates keep insisting everything's fine because they've "gotten used to it."

The Thermodynamics of Getting Uncomfortably Sweaty

The science behind heat waves is relatively straightforward, even if their impacts are anything but. As greenhouse gases trap more heat in the atmosphere, the entire climate system warms up, creating more frequent, more intense, and longer-lasting heat events. Think of it as turning up the setting on your oven—everything inside gets hotter, and it stays hot longer even after you adjust the dial (IPCC, 2021).

Several climate factors can amplify heat waves beyond just the background warming:

1. **Soil moisture depletion**: When soils dry out (which happens more readily in a warmer climate), more solar energy goes directly into heating the air rather than evaporating moisture. It's like how your sweaty body cools down through evaporation, but once you're completely dehydrated, your cooling system fails and your temperature spikes dangerously. Earth's surface works the same way.

2. **Atmospheric circulation changes**: The jet stream—that river of fast-moving air high in the atmosphere—is becoming wavier and more prone to getting stuck in place as the Arctic warms faster than the rest of the planet. When these atmospheric

patterns stall, heat can build up day after day over the same region. It's like when you're stuck in traffic behind someone who decided to have a full phone conversation before noticing the light changed—the longer the blockage persists, the more frustrating the situation becomes (NASA, 2024).

3. **Urban heat island effect**: Cities trap and hold heat thanks to concrete, asphalt, buildings, and limited vegetation. While this isn't caused by climate change, it interacts with climate change to create even more extreme heating in urban areas. The concrete jungle becomes a literal heat jungle, like taking your already hot day and adding a 400-degree oven nearby just for fun (NEEF, n.d.).

When "Record-Breaking" Becomes "Record-Shattering"

Recent years have seen heat waves that don't just break previous records —they obliterate them. The IPCC has concluded that it is an "established fact" that human-caused emissions of greenhouse gases "have led to an increased frequency and/or intensity of some weather and climate extremes" (Carbon Brief, 2022). The term "established fact" represents the highest level of certainty in IPCC terminology, even stronger than "virtually certain," which represents 99-100% probability.

Attribution studies—analyses that determine how much climate change influenced a specific weather event—have found that many recent extreme heat events would have been virtually impossible without human-caused climate change. A 2021 study found that climate change already had an impact on extreme heat events, with scientists estimating that approximately 37% of warm-season heat-related deaths between

1991 and 2018 could be attributed to anthropogenic climate change (Vicedo-Cabrera et al., 2021).

In fact, NOAA scientists with the Bulletin of the American Meteorological Society found that human-caused climate change "very likely increased the severity of heat waves that plagued India, Pakistan, Europe, East Africa, East Asia, and Australia" (NOAA, 2016). These findings highlight that the impact of climate change on heat waves is no longer theoretical but a measurable reality.

The Human Toll of Too Much Heat

Heat waves aren't just uncomfortable—they're deadly. In fact, they kill more people in an average year than any other weather-related disaster in many countries. And unlike hurricanes or floods, which make for dramatic news footage, heat waves are sometimes called "silent killers" because their victims often die quietly at home, their deaths not immediately recognized as heat-related (Center for Climate and Energy Solutions, 2023).

The human body has a fairly narrow temperature range in which it can function properly. When exposed to extreme heat, especially with high humidity, our cooling mechanisms (primarily sweating) can become overwhelmed. This leads to heat exhaustion and potentially heat stroke, which can be fatal if not treated quickly. It's like pushing your car engine into the red zone on the temperature gauge—beyond a certain point, critical systems begin to fail.

Vulnerability to heat isn't distributed equally. The most at-risk populations include:

- Older adults, whose bodies can't regulate temperature as efficiently
- Young children, whose temperature regulation isn't fully developed
- People with chronic medical conditions, particularly cardiovascular disease
- Outdoor workers exposed to direct sun and high temperatures
- Low-income communities with limited access to air conditioning or cooling centers
- Urban residents experiencing amplified heat due to the urban heat island effect (WHO, 2024)

The 2003 European heat wave resulted in approximately 70,000 excess deaths. In 2022, an analysis found that approximately 61,000 people died due to heat during Europe's hottest summer on record (WHO, 2024). These numbers exceed the death tolls from more visibly devastating disasters like hurricanes and tsunamis, yet they receive far less media attention. It's like the difference between a dramatic car crash and a slow-moving health crisis—one grabs headlines while the other quietly accumulates victims.

When It Rains, It Seriously Pours: Extreme Precipitation Events

If heat waves are the most direct consequence of global warming, extreme precipitation events are a close second. The IPCC's Sixth Assessment Report concludes that "heavy precipitation will generally become more frequent and more intense with additional global warming" (IPCC, 2021). This increased potential for extreme downpours is like an absurdly oversized water balloon waiting to burst.

The Physics of Getting Drenched

The relationship between warming and extreme precipitation follows a relatively simple principle from physics called the Clausius-Clapeyron relation. This intimidating-sounding principle (named after the two scientists who formulated it, not a villainous law firm from a John Grisham novel) describes how the water-holding capacity of air increases exponentially with temperature.

Here's what this means in practice: When it rains in today's warmer climate, the potential for extreme downpours is significantly higher than it was a few decades ago. At the global scale, the intensification of heavy precipitation is expected to follow the rate of increase in the maximum amount of moisture that the atmosphere can hold as it warms— approximately 7% per 1°C of global warming (IPCC, 2021). It's like upgrading from a modestly sized water gun to a full-on fire hose—the basic mechanism is the same, but the potential volume of water delivery has increased dramatically.

Several factors make this situation particularly concerning:

1. **Atmospheric rivers**: These narrow corridors of concentrated moisture in the atmosphere can carry staggering amounts of water. They've always existed, but now they're carrying even more moisture, leading to more extreme precipitation when they make landfall. It's like the difference between getting splashed by a passing car and being hit by a fire hydrant.

2. **Slower-moving storm systems**: Remember the wavier jet stream we mentioned earlier? This can cause weather systems to move more slowly, meaning they dump their increased moisture load over the same area for longer periods. Imagine not just a

bigger water balloon but one that follows you around all day, continuously bursting over your head (EPA, 2024).

3. **Intensified convective storms**: The increased energy in a warmer atmosphere can lead to more powerful thunderstorms and convective systems. These storm cells can rapidly dump extraordinary amounts of rain in a short time, overwhelming drainage systems designed for a previous, more moderate climate. It's like replacing your kitchen faucet with an industrial pressure washer and wondering why your sink is overflowing (EPA, 2025).

From Flash Floods to Inland Tsunamis

The real-world manifestations of these precipitation changes range from annoying to catastrophic. Flash floods—rapid flooding that occurs within six hours of heavy rainfall—are becoming more frequent and more extreme. Urban areas are particularly vulnerable due to their extensive impermeable surfaces like roads and parking lots that prevent water absorption. When extreme rain hits a city, it's like trying to drain your bathtub through a coffee stirrer—the system gets overwhelmed almost immediately (EPA, 2024).

Flooding does not solely depend on the amount of precipitation. The IPCC notes that "floods are a complex interplay of hydrology, climate and human management" (Carbon Brief, 2022). In addition to the amount and intensity of precipitation, other factors play an important role, including soil moisture, seasonal snow cover, land use, and river and catchment engineering. This means that "there is not always a one-to-one correspondence between an extreme precipitation event and a flood event, or between changes in extreme precipitation and changes in floods" (Carbon Brief, 2022).

Still, the trend is clear: EPA data shows that nine of the top ten years for extreme one-day precipitation events in the United States have occurred since 1995 (EPA, 2025). This increased frequency of extreme precipitation events translates to higher flood risks, as the IPCC projects that "a larger fraction of land areas to be affected by an increase in river floods than by a decrease in river floods" (Carbon Brief, 2022).

What makes these extreme precipitation events particularly dangerous is their unpredictability. You can't simply apply a linear scaling factor to historical rainfall patterns to predict future extremes. Instead, we're seeing complex interactions between various climate factors that can produce rainfall intensities far beyond historical experience. It's like trying to predict the behavior of a toddler who's consumed an unknown quantity of sugar—general trends might be apparent, but the specific manifestations can be surprisingly extreme.

Firestorms: When Forests Become Tinder

While water-related disasters grab headlines during the wet season, the flip side of the climate coin is equally concerning: fire. Wildfire season is getting longer, fires are burning more intensely, and they're occurring in places that historically saw few fires. It's like watching your typically calm, risk-averse friend suddenly develop an obsession with extreme sports—something fundamental has changed, and not for the better (NASA, 2025).

The Perfect Storm of Fuel, Ignition, and Meteorology

Wildfires require three elements to start and spread: fuel (vegetation), an ignition source, and favorable weather conditions. Climate change is affecting all three:

1. **Fuel conditions**: Hotter temperatures and altered precipitation patterns are creating drier vegetation, turning living, moisture-filled plants into ready-to-burn kindling. Even in areas that see increased annual precipitation, the timing matters—if rain is concentrated in shorter, more intense bursts with longer dry periods between, vegetation can still become dangerously dry. It's like how sporadic binge drinking doesn't keep you hydrated, even if your total fluid intake is technically adequate.

2. **Weather conditions**: Higher temperatures, lower humidity, and stronger winds—all influenced by climate change—create more favorable conditions for fire spread. The atmospheric vapor pressure deficit (VPD), a measure of how much moisture the air can pull from plants, has been increasing dramatically in fire-prone regions. It's essentially the atmospheric equivalent of putting a hair dryer on high and pointing it at already-dry vegetation.

3. **Ignition patterns**: While climate change doesn't directly create more lightning or human-caused ignitions, it does make any spark more likely to develop into a major fire due to the drier conditions. It's like how the ease of starting a campfire changes dramatically between using damp wood after a rainstorm versus bone-dry timber during a drought (EPA, 2025).

The combination of these factors has led to what some scientists call a "new fire era" or even the "Pyrocene"—a period of Earth's history characterized by unprecedented fire activity linked to human-caused climate change. In many regions, the concept of a traditional "fire season" is becoming obsolete as conditions conducive to major fires extend throughout much of the year.

When "Unprecedented" Becomes the New Normal

Recent years have seen truly extraordinary fire behavior that would have been considered virtually impossible under historical climate conditions. Multiple studies have found that climate change has already led to an increase in wildfire season length, wildfire frequency, and burned area (EPA, 2025). By looking back at 35 years of weather data, U.S. Forest Service scientists found that fire seasons are starting earlier in the spring and extending later into autumn. Parts of the Western United States, Mexico, Brazil, and East Africa now have fire seasons that are more than a month longer than they were 35 years ago (NASA, 2025).

A 2016 study found climate change enhanced the drying of organic matter and doubled the number of large fires between 1984 and 2015 in the western United States. A 2021 study supported by NOAA concluded that climate change has been the main driver of the increase in fire weather in the western United States (NOAA, n.d.).

What's particularly concerning is the emergence of self-reinforcing "fire weather" that creates its own meteorological conditions once fires reach a certain size. Large, intense fires can generate their own winds, create pyrocumulonimbus clouds that produce lightning (igniting new fires), and even form fire tornadoes with winds exceeding 140 mph. These phenomena represent fire behavior beyond what most firefighting strategies were designed to handle—it's like bringing water balloons to fight a volcanic eruption (Center for Climate and Energy Solutions, 2023).

Cycles of Burn, Breathe, Repeat

The impacts of increasingly severe wildfires extend far beyond the immediate burn areas. Smoke from major fires can travel thousands of miles, affecting air quality for millions of people who may live nowhere near the actual flames. During the 2020 western U.S. fire season, cities like Portland, Oregon, and Seattle, Washington, experienced air quality index (AQI) readings above 500—classified as "hazardous" and beyond the scale's intended range. In parts of California, the sky turned an apocalyptic orange as smoke particles filtered the sunlight (NASA, 2025).

This smoke contains fine particulate matter (PM2.5) that can penetrate deep into the lungs and even enter the bloodstream, causing or exacerbating respiratory and cardiovascular problems. Studies have linked wildfire smoke exposure to increased emergency room visits, hospitalizations, and mortality, particularly among vulnerable populations like those with asthma or COPD. It's like everyone in affected regions is suddenly forced to become a pack-a-day smoker, whether they choose to or not.

Moreover, the carbon released by these massive fires creates a troubling feedback loop. Forests traditionally serve as carbon sinks, absorbing CO_2 from the atmosphere. But when they burn catastrophically, they rapidly release that stored carbon, potentially shifting from net carbon sinks to carbon sources. Researchers found that carbon emissions from forest fires increased by a staggering 60% globally between 2001 and

2023 (NASA, 2025). It's like discovering your financial advisor has been secretly gambling away your savings while promising to grow your nest egg—an unpleasant reversal of what you were counting on.

When Wind Gets Weaponized: Tropical Cyclones and Tornadoes

If heat waves, floods, and wildfires weren't enough to worry about, climate change is also affecting some of nature's most violent wind events: tropical cyclones (hurricanes, typhoons, and cyclones depending on the ocean basin) and tornadoes. The relationship here is complex— these aren't simply becoming more frequent across the board—but the changes are concerning nonetheless (IPCC, 2021).

Tropical Cyclones: Fewer But Fiercer?

Contrary to what you might expect, climate models don't consistently project an increase in the total number of tropical cyclones globally. Some regions might even see fewer storms overall. However—and this is a big however—the proportion of high-intensity storms (Category 4 and 5) is increasing, and these most powerful storms are responsible for a disproportionate amount of damage (Carbon Brief, 2022).

Several climate-related factors are contributing to this intensification:

1. **Warmer ocean waters**: Tropical cyclones draw their energy from warm ocean surfaces. As climate change warms the oceans —particularly at depth, providing more total heat content— storms have access to more energy. It's like upgrading the fuel in your car from regular to premium, except instead of better engine performance, you get more catastrophic wind damage.

2. **More atmospheric moisture**: Remember that whole "warmer air holds more moisture" thing? That applies to hurricanes too, leading to higher rainfall totals from these storms. Hurricane Harvey in 2017 dumped over 60 inches of rain on parts of Texas—an amount that would have been implausible in a pre-climate change world. It's like taking a car wash and adding the pressure and volume of a fire hydrant.

3. **Slower storm movement**: Changes in atmospheric circulation patterns may be causing hurricanes to move more slowly, allowing them to dump more rain and generate more storm surge over a given area. The AR6 report notes that "it is likely that TC translation speed has slowed over the US since 1900" (Carbon Brief, 2022). It's like the difference between quickly running past a lawn sprinkler versus standing directly over it for an extended period—the total exposure makes all the difference.

These changes are already observable in the historical record. The proportion of Category 4 and 5 hurricanes has increased since the 1980s, and the average tropical cyclone intensity has increased globally. Studies of hurricane rapid intensification—when storms strengthen by at least 35 mph in 24 hours or less—show this phenomenon becoming more common, making storms harder to predict and prepare for. It's like trying to defend against a boxer who not only hits harder than before but also gives you less warning before throwing a punch.

Storm Surge: When the Ocean Attacks

One of the most destructive aspects of tropical cyclones isn't the wind or even the rain—it's storm surge, the abnormal rise in seawater level caused by a storm's winds pushing water toward shore. As the climate-enhanced hurricane's stronger winds push more water, and that water rides on top of already-higher sea levels due to global warming, the flooding potential increases dramatically. It's like the difference between getting splashed by a small wave at the beach versus being hit by a tsunami—the base mechanics are similar, but the scale and consequences are entirely different (EPA, 2016).

As relative sea level rises due to climate change, coastal flooding is becoming more frequent along much of the U.S. coastline. Most sites measured have experienced an increase in coastal flooding since the 1950s. Over the last decade, Hilo, Hawai'i, has exceeded the flood threshold most often—an average of 18 days per year—followed by Galveston, Texas, and Sewells Point, Virginia. At more than half of the locations shown, floods are now at least five times more common than they were in the 1950s (EPA, 2016).

What makes this trend particularly concerning is that coastal populations continue to grow globally, putting more people and infrastructure in harm's way just as the hazard itself is intensifying. It's like watching more and more people set up picnic blankets directly in front of a dam with visible cracks—you know it's not going to end well, but the warning signs aren't being heeded.

Tornadoes and Thunderstorms: The Wild Cards

Compared to large-scale phenomena like heat waves and tropical cyclones, the relationship between climate change and smaller, more localized severe weather events like tornadoes and severe thunderstorms is less certain. These events are too small to be directly represented in global climate models and have less consistent historical records, making trends harder to identify (IPCC, 2021).

However, research suggests that climate change is altering the environments in which severe thunderstorms and tornadoes form. The frequency of days with atmospheric conditions favorable for severe thunderstorms appears to be increasing in many regions, particularly in the eastern United States. Additionally, there's evidence that the "tornado alley" of the central U.S. may be shifting eastward over time, bringing these dangerous storms to areas with higher population densities and less historical experience with such events (IPCC, 2021).

Perhaps the most concerning change is in tornado "outbreaks"— clusters of multiple tornadoes from the same weather system. The number of tornadoes in these outbreak events appears to be increasing, meaning that when tornadic conditions develop, they're more likely to produce numerous tornadoes rather than just one or two. It's like going from occasional minor pest problems to full-scale infestations—the fundamental issue is similar, but the scale and consequences are dramatically different.

Weather Whiplash: From Drought to Deluge and Back Again
As if more intense individual weather events weren't challenging enough, climate change is also increasing the frequency of rapid transitions between opposite weather extremes—a phenomenon sometimes called "weather whiplash." These transitions from one extreme to another can be particularly damaging because natural and human systems typically have little time to recover or adapt between events (IPCC, 2021).

The Drought-Flood Cycle on Steroids

One of the most common forms of weather whiplash is the transition from extreme drought to extreme flooding. With a warming climate, this pattern is becoming more common as climate change intensifies the water cycle. Warmer temperatures increase evaporation rates, drying out soils and vegetation more quickly during dry periods. But that same warmer atmosphere can hold more moisture, leading to more intense rainfall when conditions finally align for precipitation. It's like alternating between extreme dehydration and water intoxication—neither state is healthy, and the rapid switching between them adds its own stress (Carbon Brief, 2022).

What makes these transitions particularly problematic is how they compound each other's impacts:

- Drought-hardened soils are less able to absorb rainfall, increasing runoff and flood risk
- Drought-stressed or burned vegetation provides less protection against erosion during heavy rains
- Flood-saturated soils are more vulnerable to landslides and subsequent erosion
- Infrastructure stressed by one extreme (like cracked foundations from drought) is more likely to fail when exposed to the opposite extreme

It's like how your body might handle either a fast or a feast reasonably well if given time to adjust, but switching rapidly between starvation and gluttony creates health problems beyond what either condition would cause alone.

Heat-Cold Whiplash: Temperature Rollercoasters

Another form of weather whiplash involves rapid swings between temperature extremes. Climate scientists increasingly recognize that the warming Arctic and resulting changes to the jet stream may be increasing the likelihood of these temperature whiplash events, particularly in mid-latitude regions like North America and Europe. A wavier, more meandering jet stream can allow Arctic air to plunge unusually far south, creating extreme cold events even in a warming climate. But these cold events are typically shorter-lived and bookended by periods of unusual warmth that align with the longer-term warming trend (NASA, 2024).

These temperature swings are particularly challenging for:

- **Infrastructure**: Materials expand and contract with temperature changes, and rapid shifts can accelerate wear and failure
- **Agriculture**: Plants adapted to specific growing conditions can be damaged by false spring scenarios followed by freezes
- **Ecosystems**: Wildlife cycles timed to historical temperature patterns can be disrupted
- **Human health**: Temperature volatility is associated with increased cardiovascular stress and respiratory problems

It's like trying to adjust your home thermostat that swings wildly between sauna-like heat and freezer-like cold with no comfortable middle ground—except you can't simply call a repair technician to fix Earth's climate control system.

Compound Disasters: When It All Goes Wrong at Once

Perhaps the most concerning trend in our climate-changed world is the increasing frequency of compound disasters—multiple extreme events occurring simultaneously or in close succession. These compound events often have impacts far greater than the sum of individual disasters because they overwhelm response systems and leave little opportunity for recovery between crises (IPCC, 2021).

The Disaster Multiplication Effect

When multiple extremes coincide, their impacts can multiply rather than simply add together. For example:

- A heat wave during a drought intensifies both water stress and heat-related health impacts
- Wildfires followed by heavy rain create ideal conditions for catastrophic mudslides and flash flooding
- Tropical cyclones hitting during king tides (exceptionally high tides) produce more extreme coastal flooding
- Concurrent disasters in different regions strain national emergency response resources

It's like the difference between having your car break down or losing your wallet—either one is manageable alone, but having both happen simultaneously creates a situation far more difficult than dealing with each problem independently.

The IPCC's Sixth Assessment Report concludes with high confidence that "concurrent heatwaves and droughts are becoming more frequent under enhanced greenhouse gas forcing" (IPCC, 2021). They also found medium confidence that "fire weather, i.e. compound hot, dry, and windy events, have become more frequent in some regions" and that "compound flooding risk has increased" (IPCC, 2021).

The 2020 compound disasters in the western United States illustrated this pattern vividly. Record-breaking wildfires coincided with an intense heat wave and the ongoing COVID-19 pandemic, creating extraordinary challenges for emergency management. Evacuation centers had to implement pandemic protocols, outdoor workers faced both smoke and heat stress, and healthcare systems already strained by COVID-19 had to simultaneously address heat- and smoke-related illnesses. It was the disaster equivalent of playing several difficult video games simultaneously while blindfolded—nearly impossible to manage effectively no matter how skilled the players.

The New Reality of Cascading Impacts

Beyond simple co-occurrence, disasters can trigger cascading failures across interconnected systems. These cascade sequences can transform regional disasters into national or even global crises that affect people far from the initial impact zone.

For example:

1. Heat waves stress electrical grids through increased cooling demand
2. Grid failures disable water pumping and treatment systems
3. Water system failures compound heat-related health emergencies
4. Healthcare system overload leads to increased mortality
5. Economic disruption from all of the above amplifies socioeconomic inequities

Each step in this cascade intensifies the overall impact, even for people who might have been able to manage any single element of the crisis. It's like how a minor traffic accident can trigger a massive highway pileup under the right conditions—the initial event isn't the main problem; it's the chain reaction that follows.

Climate change is increasing both the frequency of potential trigger events and the vulnerability of interconnected systems, making these cascading disasters more likely and more severe. As infrastructure ages, population density increases, and supply chains stretch across the globe, our societal resilience to these compound events may actually be decreasing just as their likelihood increases. It's like simultaneously weakening your immune system while increasing your exposure to pathogens—a concerning combination by any measure.

Attribution Science: CSI for Weather Disasters

Whenever an extreme weather event occurs, one of the first questions people ask is: "Was this caused by climate change?" This is a bit like asking whether a specific home run was caused by a baseball player's steroid use. You can't attribute any single hit directly to performance-enhancing drugs, but you can demonstrate that they increase the likelihood and magnitude of home runs overall.

The science of extreme event attribution has developed rapidly to address exactly this question. Rather than making simplistic yes/no declarations about climate change "causing" specific events, attribution studies quantify how climate change has affected the likelihood and intensity of particular types of extreme weather (IPCC, 2021).

The Science of Climate Fingerprinting

Attribution scientists use several approaches to connect extreme events to climate change:

1. **Observational analysis**: Comparing the current frequency and intensity of extreme events to historical records to identify trends and patterns
2. **Model simulations**: Running climate models with and without human-caused greenhouse gas emissions to see how the probability of specific events differs
3. **Statistical analysis**: Using statistical methods to determine whether observed changes exceed what would be expected from natural variability alone

The results are typically expressed as changes in probability or intensity —for example, "Climate change made this heat wave 30 times more likely" or "This extreme rainfall event was 20% more intense due to climate change." It's like comparing your odds of getting a parking ticket in a no-parking zone versus a legal spot—one scenario drastically increases your chances of an unpleasant outcome.

From "Can't Say" to "Virtually Certain"

The field of attribution science has progressed remarkably over the past decade. Early studies were limited to analyzing relatively simple events like heat waves, but today's attribution analyses can evaluate complex events like tropical cyclones, compound floods, and fire weather. Methods have become more sophisticated, computing power has increased, and the underlying climate science has advanced significantly (Carbon Brief, 2022).

This progress means that what was once a hesitant "we can't attribute single events to climate change" has evolved into confident statements about specific events. For example, the World Weather Attribution initiative found that the 2019 European heat wave that set all-time temperature records in countries including France, Germany, the Netherlands, and the UK was made 10-100 times more likely by climate change. Without human influence on the climate, such extreme temperatures would have been nearly impossible (NOAA, 2016).

Similarly, attribution studies of the 2020 Australian bushfires found that climate change increased the risk of the extreme fire weather conditions by at least 30%. For some recent extreme heat events in the Pacific Northwest and South Asia, the attribution conclusion was even stronger —these events would have been "virtually impossible" without human-caused climate change (IPCC, 2021).

The science has progressed to the point where attribution studies can now be conducted rapidly—sometimes within days of an event— providing timely information while the impacts are still being felt and discussed. This rapid analysis helps connect the abstract concept of climate change to concrete events that people are experiencing directly, potentially increasing public understanding and support for climate action. It's like the difference between telling someone smoking is bad for their health in general versus explaining how it contributed to their current pneumonia—the immediate connection makes the message more powerful.

Living in the New Abnormal: Adaptation Strategies

While the trends we've discussed are alarming, humans are remarkably adaptable. Throughout history, we've developed strategies to live with environmental challenges—from desert heat to Arctic cold, from flood plains to earthquake zones. The difference now is the rate of change and the fact that historical experience is becoming less reliable as a guide to future conditions. It's like trying to play a game where the rules keep changing while you're playing.

Infrastructure That Bends Without Breaking

Traditional infrastructure was designed for a relatively stable climate based on historical weather patterns. Stationarity—the assumption that future conditions will resemble the past—was a fundamental principle of engineering design. Today, that assumption is no longer valid. The 100-year flood might now occur every 20 years. The once-in-a-generation heat wave might happen annually. The historical fire season might extend to most of the year.

This new reality requires flexible, adaptive infrastructure approaches:

- **Heat-resilient cities**: Urban cooling strategies like increased tree canopy, reflective surfaces, cooling centers, and redesigned buildings can reduce urban heat island effects and protect vulnerable populations during heat waves. It's like giving our cities a permanent sun hat and SPF 50 (WHO, 2024).

- **Flood-ready development**: Rather than relying solely on hard infrastructure like levees and floodwalls, many regions are implementing "room for the river" approaches that combine engineered solutions with natural flood plains, permeable surfaces, and strategic retreat from the highest-risk areas. It's acknowledging that sometimes it's better to give water somewhere safe to go rather than trying to contain it entirely (EPA, 2024).

- **Wildfire-adapted communities**: Strategies like defensible space around structures, fire-resistant building materials, strategic fuel management, and improved evacuation planning can reduce wildfire risks even as fire behavior becomes more extreme. It's like wearing fire-resistant clothing instead of highly flammable synthetic fabrics—you're still in danger, but you've improved your odds (Center for Climate and Energy Solutions, 2023).

- **Grid resilience**: Distributed energy resources, microgrids, battery storage, and improved forecasting can help electrical systems withstand extreme weather events that would have caused widespread, prolonged outages in the past. It's like upgrading from a single engine to multiple redundant power systems—if one fails, the others can compensate.

Early Warning Systems That Actually Work

As extreme weather becomes more common and more intense, effective early warning systems become increasingly vital. Fortunately, advances in weather forecasting, climate modeling, remote sensing, and communication technologies are improving our ability to predict extreme events and alert vulnerable populations (EPA, 2024).

Modern early warning systems are becoming more:

- **Accurate**: Improved models and observations allow more precise predictions of when and where extreme events will occur

- **Timely**: Longer lead times give people more opportunity to prepare and evacuate
- **Specific**: Targeted warnings for particular neighborhoods or vulnerable groups rather than blanket alerts
- **Actionable**: Clear guidance on what specific actions to take, not just information about the hazard
- **Accessible**: Multiple communication channels to reach people regardless of technology access, language, or disability

It's like the difference between a vague "storm coming, good luck" message and a specific "severe flooding expected on your street within 4 hours, evacuate to this specific location via these routes"—the latter is vastly more useful for actually saving lives and property.

From Reactive to Proactive: Anticipatory Action

The traditional disaster management approach has been largely reactive —wait for the disaster to occur, then respond and recover. As extreme events become more frequent and severe, this approach becomes increasingly insufficient. Instead, many emergency management agencies are shifting toward anticipatory action—taking pre-planned measures when forecasts indicate a high probability of an extreme event, before the disaster actually strikes (EPA, 2024).

Anticipatory actions might include:

- Pre-positioning emergency supplies in areas likely to be affected
- Releasing funds for preparedness activities when warning thresholds are crossed
- Implementing pre-agreed evacuation plans for vulnerable populations

- Activating emergency shelters before the event rather than after
- Taking protective measures for critical infrastructure

It's like the difference between waiting until you're already sick to start taking medicine versus taking preventive measures when you know you've been exposed to a virus—the proactive approach often leads to better outcomes at lower total cost.

The Path Forward: Mitigation + Adaptation

While adaptation is essential for managing the extreme weather we're already experiencing, it's not a complete solution. Without meaningful climate mitigation—reducing greenhouse gas emissions to limit further warming—we risk outpacing our adaptive capacity as extremes become increasingly severe.

The relationship between mitigation and adaptation is sometimes framed as a choice, but they're actually complementary strategies:

- **Mitigation** reduces the rate and magnitude of climate change, keeping extremes within a manageable range
- **Adaptation** increases our resilience to the climate changes that are unavoidable due to past and current emissions

It's like dealing with a leaking roof—you both patch the hole to prevent more water from getting in (mitigation) and set up buckets to catch the water that's already coming through (adaptation). Focusing on only one approach guarantees a wet floor.

Some strategies even serve both purposes simultaneously. For example:

- Urban forests reduce urban heat islands (adaptation) while sequestering carbon (mitigation)

- Wetland restoration reduces flood risks (adaptation) while storing carbon in soils (mitigation)
- Distributed renewable energy systems provide resilience during outages (adaptation) while reducing emissions (mitigation)

These win-win approaches are particularly valuable as they address multiple aspects of the climate challenge (IPCC, 2021).

Conclusion: Disaster Isn't Destiny

The trends in extreme weather are undeniably concerning. Heat waves are becoming more frequent and intense. Precipitation patterns are shifting toward more intense events. Wildfire behavior is becoming more extreme. Tropical cyclones are intensifying. Compound and cascading disasters are increasing. If that sounds like a litany of doom, well... it kind of is.

But disaster isn't destiny. The future of extreme weather depends largely on our choices—both how much we reduce emissions to limit further warming and how effectively we adapt to the changes already underway. It's like being on a roller coaster that's approaching a steep drop—the ride is going to get more intense no matter what, but we still have some control over just how extreme it gets and how well prepared we are for the plunge.

The science is clear that differences in climate outcomes between low-emission and high-emission scenarios become increasingly dramatic after mid-century. Under high-emission scenarios, extreme events could become so severe and frequent that they overwhelm our adaptive capacity in many regions. Under low-emission scenarios that limit warming to around 2°C, extremes would still increase but likely remain within the realm where adaptation remains viable (IPCC, 2021).

So while we can't fully prevent an increase in extreme weather—that ship has already sailed with the greenhouse gases we've already emitted —we can still determine whether our future involves challenging but manageable extremes or truly catastrophic ones. It's like the difference between a difficult hiking trail with proper equipment versus attempting to scale Mount Everest in flip-flops—both are challenging, but one is potentially survivable while the other is, well, not.

In the next chapter, we'll explore another crucial aspect of our climate challenge: the energy transition that must underpin any serious mitigation effort. As we'll see, dramatic energy transformations aren't new—they've happened throughout human history. The question isn't whether our energy system will change, but how quickly and through what combination of planning versus panic. Spoiler alert: planning is definitely the better option.

Chapter 7: Energy: We've Been Here Before

Switching power sources is kinda our thing

Picture this: It's 1890, and you're standing on a street corner in New York City. Horse-drawn carriages clatter by, the air thick with the unmistakable aroma of equine emissions—about 2.5 million pounds of horse manure produced daily in the city alone. Trying not to step in the ubiquitous "road apples," you might wonder how civilization could possibly function under the crushing weight of this filthy, disease-spreading waste problem.

Fast forward just 30 years, and the streets have transformed. Automobiles have largely replaced horses, solving the manure crisis that once seemed insurmountable. Of course, those automobiles brought their own set of problems that we're still grappling with today, but the point remains: humanity has a long history of dramatically changing its primary energy sources when necessity or opportunity arise.

The transition from fossil fuels to renewable energy isn't the first energy revolution we've undergone---it's just the latest in a series of transformative shifts that have defined human development. So before you get too overwhelmed by the scale of our current energy challenge, take heart: we've been here before. Different stakes, different technologies, but the same fundamental process of reinventing how we power our world.

In this chapter, we'll explore previous energy transitions, examine what drives these shifts, and see what lessons they might offer for our current renewable revolution. Spoiler alert: energy transitions are never easy, never happen overnight, and always create both winners and losers. But they do happen, and they've generally moved humanity forward---even if the road gets a bit bumpy along the way.

The Original Energy Transition: Fire and Agriculture

Long before fossil fuels entered the picture, humans underwent their first major energy transition when our ancestors figured out how to control fire roughly 400,000 years ago. This wasn't just a neat party trick---it was a revolution in human energy use that fundamentally changed our species' trajectory.

Fire allowed early humans to:

- Cook food, making more nutrients bioavailable and reducing the energy needed for digestion
- Stay warm in colder climates, expanding our habitat range
- Keep predators at bay, improving survival rates
- See in the dark, extending productive hours
- Transform materials like clay into pottery and later, metals into tools

In essence, fire gave humans our first external energy source beyond muscle power, dramatically increasing our energy access. It was like upgrading from a basic smartphone battery that dies by noon to one of those portable power banks that can charge your device multiple times over. The energy return on investment (EROI)---how much energy you get out compared to how much you put in---was enormous for a species previously limited to whatever calories they could digest.

The next major energy shift came with the agricultural revolution about 10,000 years ago. By domesticating plants and animals, humans harnessed solar energy more efficiently through crops and livestock.

This was essentially solar power with biological batteries---plants captured sunlight through photosynthesis, and humans either ate those plants directly or fed them to animals that we then consumed (Smil, 2017).

Agriculture allowed for energy storage (grain silos, preserved food, livestock on the hoof) and created energy surpluses that freed some people from food production to specialize in other areas. Without these early energy transitions, we'd still be nomadic hunter-gatherers with limited population density and technological development. It was the energy equivalent of going from living paycheck to paycheck to having a savings account and investment portfolio.

From Water and Wind to Coal: The First Industrial Revolution

For thousands of years after agriculture took hold, human civilization relied primarily on renewable energy flows---human and animal muscle, biomass burning, and the mechanical power of water and wind. Waterwheels appeared as early as the 3rd century BCE, while windmills became common in Persia by the 7th century CE and spread throughout Europe during the Middle Ages.

These pre-industrial renewables were surprisingly powerful. By the 18th century, the Netherlands had over 10,000 windmills generating about 20 watts of power per capita---comparable to what many developing countries had in the early 20th century. Water power was even more significant, with mills generating power equivalent to millions of human laborers (The Conversation, 2024).

But these energy sources had limitations:

- **Geographic constraints**: You needed to be near moving water or in windy locations
- **Intermittency**: Output varied with weather and seasons
- **Power density**: Energy was relatively diffuse and difficult to concentrate

Enter coal---the dense, portable, energy-rich fuel that would power the Industrial Revolution. Though coal had been used for heating since ancient times, its transformation into mechanical energy through the steam engine unleashed unprecedented economic change.

The shift from water and wind to coal-powered steam wasn't driven by resource scarcity (rivers weren't running out) but by the superior characteristics of the new energy source: coal could be transported to factories rather than building factories by rivers, steam engines could operate 24/7 regardless of weather, and the power density was dramatically higher. As energy historian Vaclav Smil points out, "every transition to a new energy supply has to be powered by the intensive deployment of existing energies and prime movers" (The Conversation, 2024).

This transition was messy, contested, and uneven. The air in industrial cities became thick with coal smoke, working conditions in early factories were often brutal, and traditional craftspeople found their livelihoods threatened by mechanization. Sound familiar? Energy transitions typically feature early adoption by commercial interests seeking competitive advantage, followed by broader social changes as the new systems reshape society---not unlike what we're seeing with renewable energy today.

From Coal to Oil: Liquefying the Energy System

By the late 19th century, coal had transformed global industry, transportation, and urban development. It seemed the energy future was solidly carboniferous. Then came oil.

First used primarily for lighting (kerosene lamps), petroleum's true potential emerged with the internal combustion engine and the automobile. Oil offered significant advantages over coal:

- Higher energy density (more energy per pound)
- Easier to transport via pipelines
- Cleaner burning at the point of use
- More convenient to store and handle
- Better suited for mobile applications

The transition from coal to oil dominance wasn't immediate or complete. Coal remained king in electricity generation and some industrial processes, while oil captured transportation and gradually expanded into other sectors. This hybrid pattern---where new energy sources first dominate in applications where they have clear advantages, then gradually expand---offers valuable insight for our current transition (World Economic Forum, 2022).

Oil transformed not just energy systems but geopolitics and social organization. Nations with oil reserves gained new strategic importance, urban development patterns shifted with mass automobile adoption, and new industries emerged around petrochemicals. The transition created tremendous wealth while also embedding fossil fuels ever more deeply into the foundations of modern life---a challenge today's energy transition must overcome.

What's notable about this shift is that it wasn't motivated by climate concerns or even local pollution (though oil was generally cleaner than coal at the point of use). It was driven by oil's superior technical and economic characteristics for key applications. The market largely led the transition, with policy following rather than leading---a pattern that differs somewhat from today's renewable transition, where policy has played a more central role, at least initially (Dahl 7).

The Electrification Revolution: From Direct Power to Energy Carriers

The most transformative energy shift of the 20th century wasn't about primary energy sources but about energy carriers and end-use technologies: electrification. Before electricity, energy had to be used near where it was generated---mechanical power from a waterwheel turned a millstone directly, coal boiled water to create steam that directly drove factory equipment.

Electricity changed everything by separating energy generation from energy use. Power could be produced at distant plants, converted to electricity, transmitted over wires, and reconverted to useful forms (light, heat, motion) at the point of use. This flexibility unleashed incredible innovation in both energy production and consumption technologies (RMI, 2025).

The electrification process was remarkably gradual by today's standards. Thomas Edison opened the first commercial power plant in 1882, but by 1925---over 40 years later---only half of U.S. homes had electric service. Rural electrification came even later, requiring significant government intervention through the Rural Electrification Administration established in 1935.

This leisurely timeline offers both caution and comfort for our current transition: caution because we don't have 50+ years to address climate change, but comfort in knowing today's renewable transition is actually proceeding much faster than historical precedent would suggest (Visual Capitalist, 2022).

Electrification also demonstrated how energy transitions create new possibilities rather than just replacing old systems. Electric light was superior to gas lighting, electric motors enabled factory layouts impossible with steam power, and countless new devices---from refrigerators to computers---simply had no predecessor in previous energy regimes. Similarly, our current transition isn't just about replacing fossil electricity with renewable electricity, but about enabling entirely new configurations of how we produce, deliver, and use energy.

The Current Transition: Old Patterns, New Dynamics

Today's shift toward renewable energy follows many patterns seen in previous transitions while also featuring distinctive characteristics driven by climate urgency, technological change, and evolving social priorities.

1. Learning from History: What's Common Across Transitions

Historical energy transitions consistently show certain patterns:

- **They take time**: Typically 50-100 years for a significant global shift in primary energy
- **They're driven by advantages**: New energy sources succeed by being better/cheaper for key applications
- **They're geographically uneven**: Leaders forge ahead while laggards maintain older systems

- **They transform in stages**: New energy sources first dominate in areas of clear advantage before broader adoption
- **They reshape society**: Energy transitions drive changes far beyond the energy sector itself
- **They face resistance**: Incumbent interests fight to preserve their position
- **They create new winners**: Entirely new industries and fortunes emerge from the transition

These historical patterns suggest the renewable transition won't be instantaneous, universal, or without conflict. Like previous shifts, it will likely proceed as a rolling transformation rather than a single global change, with some regions switching rapidly while others maintain fossil systems longer (Smil, 2010).

2. What's Different This Time: Novel Features of the Renewable Transition

While history offers useful parallels, today's energy transformation differs from previous transitions in several important ways:

- **Climate motivation**: Unlike previous shifts driven primarily by economic and convenience factors, climate change creates urgency to move faster than market forces alone might dictate
- **Global coordination**: International agreements and targets create an unprecedented level of coordinated action across countries
- **End-of-pipe to full-lifecycle focus**: Environmental considerations now include not just point-of-use pollution but entire production chains and carbon footprints

- **Distributed potential**: Many renewable technologies enable decentralized, customer-owned generation rather than only centralized production
- **Digital integration**: Smart systems allow for unprecedented monitoring, control, and optimization of energy flows
- **Circular economy awareness**: Greater focus on material inputs and waste outputs alongside energy performance

Perhaps most significantly, renewable technologies follow different economic patterns than conventional energy---their costs decline with cumulative production rather than increasing as the best resources are exploited. Solar photovoltaic module prices have fallen by roughly 99% since 1976, wind turbine costs have declined about 70% since 2009, and battery prices have dropped nearly 90% in the past decade (IEA, 2024).

This "learning curve" effect creates the potential for a virtuous cycle and accelerating transition: greater deployment drives lower costs, which enables greater deployment, which further reduces costs. It's like if oil wells actually produced more oil the more we drilled, or if coal mines became safer and more productive the more we dug---a fundamentally different dynamic than extractive industries typically exhibit.

The Renewable Transition: Where Are We Now?

Current progress in the renewable energy transition varies dramatically by sector and region. Looking at the global picture:

1. Electricity: Leading the Charge

The power sector is furthest along in decarbonization. Renewable energy has grown from a niche contributor to a mainstream power source representing about 29% of global electricity production as of

2020. In some regions, the transformation is even more dramatic---renewables provide more than 40% of electricity in the European Union, over 50% in Denmark, and nearly 100% in Iceland and Norway.

New capacity additions tell an even more striking story: renewables accounted for about 90% of new electricity generation capacity added globally in 2020. In many markets, wind and solar are now the cheapest form of new electricity generation, even without subsidies. The economic case for renewables in the power sector has become so compelling that market forces are now accelerating rather than resisting the transition (IRENA, 2023).

Perhaps most tellingly, financial markets are increasingly betting on renewable dominance. Major power companies are reorienting their business models around renewable development, investors are demanding climate risk disclosure, and fossil fuel assets are facing growing risk of becoming stranded---worth less than their book value due to policy changes or market evolution. When the money starts moving, change typically accelerates (Climate Champions, 2023).

2. Transportation: Momentum Building

The transportation sector presents a more mixed picture. Electric vehicles represent a small but rapidly growing share of new car sales---reaching over 10% globally in 2022, with much higher percentages in leading markets like Norway (over 90%), China, and parts of Europe.

Commercial vehicles, shipping, and aviation face greater challenges due to weight, range, and power density requirements, though solutions are emerging. Electric buses are becoming common in many cities, while hydrogen, advanced biofuels, and synthetic fuels show promise for harder-to-electrify transportation segments.

The transition pace in transportation depends heavily on both technology development and infrastructure deployment---electric vehicles need charging networks, hydrogen vehicles need fueling stations, and alternative fuel aircraft need new supply chains. These interdependencies create potential chicken-and-egg problems but also opportunities for coordinated system transformation (IEA, 2023).

3. Buildings: A Renovation Challenge

Buildings consume about one-third of global energy, primarily for heating, cooling, and various appliances and equipment. The path to decarbonization involves both electrifying end uses (replacing gas furnaces with heat pumps, gas stoves with induction) and improving efficiency through better insulation, windows, and energy management systems.

Progress varies widely by region, with some countries requiring near-zero energy performance for new construction while others maintain minimal standards. The greater challenge lies in retrofitting existing buildings, as the vast majority of structures that will exist in 2050 have already been built. Accelerating the renovation rate of existing buildings from the typical 1-2% annually to 3-5% would dramatically speed the transition (IPCC, 2022).

4. Industry: The Final Frontier

Heavy industry---including steel, cement, chemicals, and manufacturing---presents some of the toughest decarbonization challenges. These sectors often require high-temperature process heat, currently supplied primarily by fossil fuels, and some industrial processes (like cement production) release CO_2 as an inherent byproduct beyond just energy-related emissions.

Solutions are emerging but remain at earlier stages than in other sectors. Options include:

- Electrification where technically feasible
- Green hydrogen from renewable electricity for high-temperature processes
- Biomass for certain applications
- Carbon capture for processes that can't eliminate emissions
- Material efficiency and circular economy approaches to reduce primary production needs

Progress in industrial decarbonization will likely require significant policy support, from carbon pricing to clean procurement requirements, alongside continued technological development (IRENA, 2024).

The Essential Elements: What's Needed to Accelerate the Transition

While the renewable transition is underway and accelerating in many regions, meeting climate goals requires further acceleration across all sectors. Historical patterns and current dynamics suggest several key elements that can speed the transformation:

1. Policy Frameworks: Beyond One-Off Interventions

Effective energy transition policies create long-term, stable frameworks rather than short-term incentives. Examples include:

- **Carbon pricing**: Putting a cost on emissions through carbon taxes or cap-and-trade systems internalizes climate impacts in economic decisions
- **Clean energy standards**: Requirements for increasing percentages of clean electricity, fuels, or materials create clear market signals
- **Performance standards**: Energy efficiency requirements for vehicles, buildings, and equipment drive continuous improvement
- **Infrastructure planning**: Coordinated approaches to grid modernization, charging networks, and other enabling infrastructure reduce barriers

These policy approaches are most effective when they provide regulatory certainty over investment timeframes---giving businesses and individuals confidence to make long-term decisions aligned with decarbonization pathways (World Economic Forum, 2023).

2. Finance: Redirecting Capital Flows

The global energy system is enormously capital-intensive, with trillions of dollars invested annually in infrastructure, equipment, and maintenance. Redirecting these financial flows toward clean energy solutions requires:

- **Risk mitigation instruments**: Loan guarantees, insurance products, and other tools that reduce perceived risks of clean energy investments
- **Green bonds and loans**: Dedicated financial products for climate-friendly projects

- **Divestment and ESG investing**: Shifting capital away from fossil-intensive industries toward cleaner alternatives
- **Blended finance**: Combining public and private capital to fund projects that might not attract purely commercial investment

Financial institutions are increasingly recognizing both climate risks and opportunities, with many major banks, insurers, and asset managers announcing climate commitments and developing sustainable finance products (Climate Champions, 2023).

3. Just Transition: The Human Dimension

Energy transitions invariably create winners and losers. While previous transitions occurred with little consideration for those displaced, the current shift features greater awareness of the need for a "just transition" that addresses the human impacts of moving away from fossil fuels (Wikipedia, 2025).

Communities built around coal mining, oil extraction, or gas production face genuine economic challenges as these industries decline. Workers in these sectors often have specialized skills that don't easily transfer to other industries, and the jobs typically offer better wages and benefits than available alternatives in their regions. It's like being a highly skilled typewriter repair technician when computers became dominant---your expertise suddenly has diminishing value through no fault of your own.

Just transition strategies include:

- Targeted workforce development and retraining programs
- Early retirement options for workers near the end of their careers
- Economic diversification investments in affected communities

- Cleanup and remediation jobs that utilize existing workforce skills
- Social safety net enhancements to support workers during transition
- Involving affected communities in transition planning rather than imposing solutions

These approaches recognize that energy transitions are not just technological projects but social transformations that affect real people's livelihoods and communities. By proactively addressing these human dimensions, we can reduce opposition to necessary changes and ensure the benefits of the clean energy economy are widely shared (IRENA, 2023).

4. Crossing Sectors: Beyond Electricity

While electricity generation has made impressive progress toward decarbonization in many regions, other sectors---transportation, buildings, industry, and agriculture---have seen slower transformation. Addressing these sectors requires both electrification where possible and developing alternative solutions for applications where direct electrification is challenging.

Industrial processes like steel and cement production, long-distance transportation like shipping and aviation, and some agricultural emissions require specialized approaches beyond simple electrification. Options include:

- Green hydrogen produced from renewable electricity for industrial processes and possibly heavy transportation

- Sustainable biofuels for applications requiring high energy density
- Advanced materials that require less energy-intensive production
- Carbon capture for processes that can't eliminate emissions entirely
- Circular economy approaches that reduce primary material needs

The transition beyond electricity will likely be more challenging and potentially slower than the power sector transformation, but the emerging portfolio of solutions suggests viable pathways exist for even the most difficult applications (IEA, 2023).

Reasons for Optimism: Better Than Previous Transitions

While the climate timeline creates unprecedented urgency, the current energy transition actually has several advantages over previous shifts that offer reasons for optimism:

1. Superior Economics Driving Deployment

Unlike previous transitions that often involved trade-offs between economic and environmental factors, renewables increasingly win on pure cost grounds. In most global markets, new solar and wind are now the cheapest forms of electricity generation---cheaper than new coal, gas, or nuclear plants, and in many cases, cheaper than continuing to operate existing fossil plants (IEA, 2024).

This economic advantage creates a powerful market pull beyond any policy push. Even in regions with limited climate policies, the pure financial case for renewables is driving rapid adoption. It's like how digital photography replaced film---not because of environmental regulations but because the new technology ultimately became cheaper, more convenient, and better performing (IRENA, 2025).

2. Multiple Co-Benefits Beyond Climate

Renewable energy offers numerous benefits beyond carbon reduction that can motivate adoption even among those unconvinced by climate concerns:

- **Energy independence**: Domestically produced renewable energy reduces dependency on imported fuels and improves national security
- **Air quality improvements**: Eliminating fossil combustion dramatically reduces harmful air pollutants that cause respiratory and cardiovascular disease
- **Water conservation**: Renewables require minimal water compared to thermal power plants
- **Job creation**: Clean energy industries typically create more jobs per unit of energy than fossil alternatives
- **Price stability**: Once built, renewables have no fuel costs and thus provide long-term price predictability

These co-benefits create broader constituencies for the energy transition and can align diverse interests behind common solutions, potentially accelerating adoption beyond what climate concern alone would achieve (IRENA, 2021).

3. Unprecedented Technological Learning Curves

Renewable technologies follow exponential improvement patterns similar to information technologies rather than the incremental progress typical of conventional energy. Solar photovoltaic costs have declined by roughly 99% since 1976, while wind costs have fallen by approximately 70% since 2009. Battery prices have plummeted by nearly 90% in the past decade.

These learning curves don't follow time-based improvements but rather production-based improvement---each doubling of cumulative production reduces costs by a predictable percentage. This means that accelerating deployment creates a virtuous cycle: more deployment leads to lower costs, which enables more deployment, which further reduces costs (Visual Capitalist, 2022).

4. Modular Scaling Advantages

Unlike conventional energy infrastructure built through massive, lumpy investments (nuclear plants, refineries, LNG terminals), renewable energy is modular---deployed in incremental units from residential rooftop solar to utility-scale wind farms. This modularity creates several strategic advantages:

- **Faster learning**: More frequent iteration cycles accelerate technological improvement
- **Lower financial risk**: Smaller incremental investments reduce the consequence of any single project failure
- **Geographic flexibility**: Deployment can occur wherever resources and demand exist, not just where massive infrastructure can be accommodated
- **Incremental addition**: Capacity can be added gradually to match demand growth rather than requiring large advance commitments

This modularity allows renewable deployment to follow patterns more like consumer electronics than traditional infrastructure, potentially enabling much faster scaling than historical energy transitions achieved (IEA, 2023).

Conclusion: We've Done This Before, But Never Quite Like This

Throughout human history, we've repeatedly transformed our energy systems---from muscles to fire, from water wheels to steam engines, from coal to oil, and from mechanical power to electricity. Each transition changed not just how we powered our activities but the very nature of those activities and the societies they enabled.

The current shift to renewable energy follows many historical patterns while featuring unique characteristics---it's both a return to harvesting natural energy flows as we did before fossil fuels and a leap into a fundamentally new relationship with energy based on advanced technology. It combines the deliberate direction of a managed transition with the economic inevitability of superior technologies displacing their predecessors.

What's clear from this historical perspective is that energy transitions do happen, have happened repeatedly throughout human development, and will continue to happen. The question isn't whether we'll transition to a predominantly renewable energy system---the economics increasingly make that outcome inevitable. The real questions are how quickly we'll make the shift, how we'll manage the human and social dimensions of the transition, and whether we'll move fast enough to meet our climate goals.

The good news is that we're not starting from scratch. We have centuries of experience with energy transitions, a growing toolkit of policy and financial mechanisms to accelerate change, and renewable technologies that offer genuine economic and performance advantages over the systems they're replacing. It's like having both a detailed roadmap from previous journeys and a significantly better vehicle for the trip.

The decisions we make now will determine whether this transition becomes another chapter in humanity's long history of energy revolutions or something more disruptive---whether we manage a controlled transformation or face a chaotic scramble as climate impacts force our hand. As we'll explore in the next chapter, the detailed evidence from tree rings and other natural archives makes clear that we don't have the luxury of a leisurely transition timeframe.

Energy transitions are challenging but normal. We've been here before, and we'll get through this one too. The question is whether we'll do so with foresight and fairness or with procrastination and pain. History suggests the former path is entirely possible if we choose it.

Chapter 8: Dead Trees Can Talk

And they're gossiping about your climate sins

Next time you're enjoying the shade of an old tree, take a moment to consider that you're standing next to the botanical equivalent of a time capsule with built-in climate recording equipment. That oak might look like it's just silently minding its business, occasionally dropping acorns on unsuspecting pedestrians for entertainment, but it's actually been meticulously documenting climate conditions for decades, even centuries. Trees are basically the world's most patient journalists, steadily recording what's happening around them while waiting for someone to finally read their work.

In this chapter, we'll explore how scientists use tree rings, ancient logs, and other woody evidence to reconstruct climate histories that stretch back hundreds, thousands, and even millions of years. It's like CSI: Climate Scene Investigation, if the star witnesses were plants and the smoking gun was actually a chainsaw. So grab your magnifying glass and detective hat—we're about to interrogate some trees about what they've witnessed.

Nature's Record Keepers: How Trees Document Climate

Trees might not seem like the most technologically advanced data recorders, but they've been perfecting their craft for about 385 million years, which is significantly longer than humans have been making climate observations. While we've only been keeping systematic weather records for about 150 years (roughly coinciding with our decision to start messing with the climate—timing!), trees have been quietly documenting environmental conditions since before dinosaurs roamed the Earth (NASA, 2017).

Tree Rings 101: Annual Growth Diaries

The basic principle behind tree rings is wonderfully simple: most trees in temperate and boreal regions add one growth ring each year. During the growing season (typically spring and summer), trees produce lighter-colored "early wood" with larger cells that efficiently transport water. As growth slows later in the season, they produce darker "late wood" with smaller, denser cells. This contrast creates the distinctive banded pattern we see when looking at a cross-section of a tree trunk—like nature's barcode (NOAA, 2018).

What makes these rings useful to climate scientists is that their characteristics—width, density, and chemical composition—vary depending on growing conditions. In good years with favorable temperature and abundant moisture, trees produce wider rings. During droughts or cold spells, growth slows, resulting in narrower rings. It's essentially the tree's autobiography, documenting the good times (wide rings) and the tough years (narrow rings) throughout its life (NASA, 2017).

Of course, trees don't just respond to climate—they also react to competition, disease, fire, and that time your grandfather carved his initials into the bark. But by carefully selecting trees from climate-sensitive locations and analyzing large samples, scientists can filter out these non-climate influences to reveal the underlying climate signal. It's like separating the climate signal from the noise, except the "noise" might include that lightning strike in 1842 or the time a beaver decided your tree looked tasty (NOAA, 2018).

Dendrochronology: The Science of Tree-Ring Dating

The scientific discipline of studying tree rings is called dendrochronology (from the Greek "dendron" meaning tree and "chronos" meaning time, not to be confused with "dendro-chronology" which would be the study of how slowly trees tell stories). This field was pioneered by astronomer A.E. Douglass in the early 20th century as he searched for connections between sunspot cycles and climate. While looking for evidence of solar influences, he stumbled upon something even more valuable—a method for precisely dating wooden archaeological artifacts and reconstructing past climate conditions (Carbon Brief, 2018).

The key insight of dendrochronology is that trees growing in the same region at the same time experience similar climate conditions and thus develop similar ring patterns. By overlapping ring sequences from living trees with those from older dead trees and archaeological wood, scientists have built continuous chronologies spanning thousands of years. It's like creating a relay race through time, where each tree passes the climate baton to an older specimen (NOAA, 2018).

Some of these chronologies are impressively long:

- The German oak-pine chronology stretches nearly 12,600 years into the past
- The bristlecone pine chronology from the western United States reaches back more than 8,000 years (NOAA Climate.gov, 2018)

These aren't just vague estimates—dendrochronology provides annual precision. When a dendrochronologist says something happened in 1242 CE, they don't mean "around the 13th century" or "give or take a decade." They mean specifically 1242—a level of precision rarely matched in paleoclimate reconstruction. It's like having a time-stamped climate receipt going back thousands of years (NOAA, 2018).

Climate Detectives: What Tree Rings Reveal

Tree rings don't just tell us when things happened—they reveal detailed information about past climate conditions. Depending on the species, location, and specific measurement, tree rings can record:

1. **Temperature**: In cold, high-elevation or high-latitude locations, tree growth is primarily limited by summer temperatures, making ring width or density in these trees an excellent temperature proxy. It's like how your motivation to go outside is primarily limited by temperature—when it's -20°F, you're not making much "personal growth" happen beyond the couch.

2. **Precipitation**: In arid or semi-arid regions where water availability limits growth, ring widths strongly correlate with precipitation amounts. These trees are essentially rain gauges with bark (NOAA, 2018).

3. **Drought severity**: By combining temperature and precipitation signals, scientists can reconstruct historical drought conditions more comprehensively than either variable alone. The trees keep score of which years were truly miserable on multiple fronts (USGS, 2022).

4. **Stream flow**: Trees growing near rivers and streams record information about water availability that can be used to reconstruct historical river flow rates. These riparian trees are nature's hydrologists, documenting floods and dry spells long before humans thought to install gauge stations (USGS, 2022).

5. **Extreme events**: Sudden growth reductions, physical damage, or anatomical changes in rings can record extreme events like fires, volcanic eruptions, insect outbreaks, or severe freezes. It's like the tree's way of saying, "Let me tell you about this CRAZY thing that happened back in 1783..." (RealClimate, 2015).

By analyzing these tree-ring archives, climate scientists have reconstructed seasonal temperature and precipitation patterns for much of the globe over the past several centuries to millennia. This long-term perspective is crucial for understanding natural climate variability and placing recent changes in context. It's like having climate security camera footage from long before actual cameras were invented (NOAA, 2018).

Beyond Rings: Other Ways Trees Talk About Climate

While rings get most of the attention, trees have other ways of documenting climate conditions. Think of tree rings as the main text of the climate story, while these other methods provide the fascinating footnotes and appendices:

Isotope Analysis: Chemical Climate Fingerprints

Trees don't just vary their growth rates with climate—they also incorporate different isotopes of elements like oxygen, carbon, and hydrogen depending on environmental conditions. An isotope is a version of an element with a different number of neutrons, making it slightly heavier or lighter than the standard version. It's like having twins who look identical but one inexplicably weighs 10 pounds more.

The ratio of these isotopes in tree rings provides information about:

- **Oxygen isotopes ($^{18}O/^{16}O$):** Reflect source water and precipitation patterns
- **Carbon isotopes ($^{13}C/^{12}C$):** Indicate water stress and photosynthetic efficiency
- **Hydrogen isotopes ($^{2}H/^{1}H$):** Record hydrological conditions and water sources

What makes isotope analysis particularly valuable is that it can provide climate information from regions where ring width alone might not be a strong climate indicator. For example, tropical trees often don't produce distinct annual rings because they grow year-round, but isotope analysis can still reveal seasonal climate variations in their wood. It's like getting climate information from trees that refused to keep an organized calendar (IPCC, 2013).

Subfossil Wood: When Dead Trees Still Speak

While living and recently dead trees provide climate information for the past few centuries to millennia, older "subfossil" wood—preserved in bogs, lakes, glaciers, or permafrost—extends these records even further back in time. These ancient wood specimens aren't fully fossilized (turned to stone) but have been preserved in oxygen-poor environments that prevent normal decomposition (Nature, 2022).

Subfossil logs have been discovered that are tens of thousands of years old, with some of the oldest well-preserved wood specimens dating to around 50,000 years ago. These ancient trees provide windows into climate conditions during the last ice age and glimpses of the world our ancestors experienced. It's like finding your great-great-great-great-grandfather's weather journal, except instead of writing "quite chilly today," he grew an extra-narrow tree ring (NOAA, 2018).

Pollen in Tree Growth: Environmental Snapshots

Trees don't just record climate information in their own tissues—they also produce pollen that gets preserved in lake sediments, bogs, and other environments. By analyzing these pollen records, scientists can reconstruct past forest compositions and climate conditions.

Changes in the types and abundances of pollen over time reflect shifts in temperature, precipitation, and growing conditions. For example, a shift from oak and hickory pollen to spruce and fir would indicate cooling conditions, while the reverse would suggest warming. It's like trees sending text messages about the climate to future scientists.

Pollen records can extend much further back than tree rings—in some locations, continuous pollen records cover hundreds of thousands of years. Combined with other proxy sources, these ancient plant text messages help fill in the climate story where tree rings alone can't reach (IPCC, 2013).

Reading the Rings: What Trees Tell Us About Climate History

Now that we understand how trees record climate information, let's explore what they've revealed about Earth's climate history. Like any good witness, trees have documented some fascinating climate episodes long before humans started keeping systematic records:

The Medieval Warm Period: Not Your Modern Warming

Tree rings provided some of the first detailed evidence for the Medieval Warm Period (roughly 950-1250 CE), a time when parts of Europe, North Atlantic regions, and some other areas experienced temperatures warmer than the subsequent centuries (though not as warm as today). This relatively balmy period coincided with Viking settlements in

Greenland, vineyards in northern England, and agricultural prosperity in many European regions (IPCC, 2013).

However, tree ring networks have also revealed something crucial about this period—it wasn't a globally synchronized warm episode like our current warming. Different regions experienced their warmest decades at different times, and some regions showed no unusual warmth at all. It was more like a climatic game of hot potato, with warmth shifting from region to region, rather than the globally coordinated warming we're experiencing now. Medieval warming was like having several small, localized fever spots, while current climate change is like the whole planet having a persistent fever (IPCC, 2013).

This distinction is important because climate skeptics sometimes point to the Medieval Warm Period as evidence that current warming might be natural. Tree rings tell a more nuanced story—yes, natural variability can cause significant regional warming, but the current globally synchronized warming with no natural explanation is a very different beast. It's like comparing a local power outage to a nationwide blackout —similar symptoms but completely different causes and implications (IPCC, 2013).

The Little Ice Age: When Europe Caught a Cold

Following the Medieval Warm Period, tree rings document a prolonged cool period known as the Little Ice Age (roughly 1300-1850 CE). During this time, mountain glaciers advanced, rivers and canals frequently froze, and crop failures and famines were more common in many regions. If you've seen paintings of people ice skating on the Thames River in London or winter festivals on frozen Dutch canals, you're looking at Little Ice Age climate conditions that rarely occur in those locations today (IPCC, 2013).

The Little Ice Age wasn't uniformly cold throughout its duration but rather featured several particularly cold periods separated by more moderate intervals. Tree rings have helped pinpoint these cold phases and identify their potential causes, including:

- Reduced solar activity during periods like the Maunder Minimum (1645-1715)
- Increased volcanic activity injecting cooling aerosols into the stratosphere
- Changes in ocean circulation patterns affecting heat distribution
- Possible feedback processes involving sea ice, ocean currents, and atmospheric patterns

Like the Medieval Warm Period, the Little Ice Age showed regional variations rather than uniform global cooling. It's as if Earth was wearing a poorly designed sweater with some areas too cold and others relatively comfortable—not the globally coordinated temperature changes we're seeing today (IPCC, 2013).

Volcanic Cooling: Trees Remember Eruptions

One of the most dramatic climate stories told by tree rings involves the cooling effects of major volcanic eruptions. When volcanoes inject sulfur aerosols into the stratosphere, they reflect incoming sunlight and cool the planet for several years. Tree rings record these cooling episodes with distinctive narrow rings or frost damage that can be precisely dated (RealClimate, 2015).

The 1815 eruption of Mount Tambora in Indonesia—the largest eruption in recorded history—led to the infamous "Year Without a Summer" in 1816, when frost and snow occurred in June and July across parts of North America and Europe, causing widespread crop failures and famine. Tree rings from this period show some of the narrowest growth rings of the past several centuries, providing physical evidence of this climate disaster. It was essentially Earth turning down the thermostat to an uncomfortable level, and trees documented the whole chilly episode (RealClimate, 2015).

By matching unusual tree ring patterns with ice core records of volcanic sulfate deposits, scientists have identified the cooling impacts of many prehistoric eruptions that occurred before written records. This volcanic detective work has revealed that massive eruptions have periodically caused global cooling of 1-2°C that lasted several years—temporary climate impacts comparable in magnitude (though opposite in direction) to the warming we're currently experiencing due to greenhouse gases. The key difference? Volcanic cooling is temporary, while our greenhouse warming will persist for centuries without emission reductions (IPCC, 2013).

Megadroughts: When Dry Spells Become Disasters

Tree rings have revealed something particularly concerning about drought history, especially in regions like the American West: megadroughts. These prolonged dry periods lasting decades were far more severe and persistent than anything experienced in the modern instrumental record. It's like discovering that what you thought was a "bad hair day" is nothing compared to the catastrophic mullet your dad sported for all of 1987 (USGS, 2022).

Reconstructions from tree rings show that the Western United States experienced multiple megadroughts during the past 1,200 years, including:

- A 240-year-long drought in the Sierra Nevada from approximately 850-1090 CE
- An intense drought in the Colorado River Basin in the mid-1100s
- Severe 16th-century drought that affected much of North America
- An exceptional 2nd century drought that reduced Colorado River flow by 32%, unmatched in severity even by the current 22-year drought (USGS, 2022)

These tree-ring-documented megadroughts far exceed the severity and duration of more recent droughts like the 1930s Dust Bowl, which devastated agriculture and forced mass migrations. They suggest that what we consider "extreme" drought based on our limited instrumental record is actually well within the range of natural variability—an unsettling revelation as climate change pushes many regions toward hotter, drier conditions that could make such megadroughts more likely (USGS, 2022).

This historical perspective from tree rings provides crucial context for water management and planning in drought-prone regions. It's like learning that the "100-year flood" level on which you based your home buying decision actually happens every 20 years—information you really would have appreciated before signing those mortgage papers (Colorado Sun, 2022).

The Hockey Stick: Why Tree Rings Made Climate Deniers Nervous

In 1998, climate scientists Michael Mann, Raymond Bradley, and Malcolm Hughes published a tree-ring-based temperature reconstruction for the Northern Hemisphere covering the past 600 years, later extended to 1,000 years. Their resulting graph—showing relatively stable temperatures for centuries followed by a sharp upward trend in the 20th century—resembled a hockey stick lying flat with the blade pointing upward at the modern end. This "hockey stick" graph became one of the most iconic and controversial images in climate science (RealClimate, 2021).

What the Hockey Stick Shows

The hockey stick reconstruction illustrated several key points:

1. Northern Hemisphere temperatures were relatively stable for centuries (with some medieval warmth and Little Ice Age cooling)
2. 20th-century warming was unprecedented in both rate and magnitude compared to the previous millennium
3. Recent temperatures exceed anything experienced during at least the past 1,000 years

This dramatic visual demonstration of unusual modern warming made the hockey stick graph a powerful communication tool—and an immediate target for those opposing climate action. It's like how your meticulously documented household budget becomes suddenly fascinating to your spouse when it clearly shows they're the one spending too much on "unnecessary" purchases (RealClimate, 2021).

Controversy and Confirmation

Few scientific visualizations have been subjected to as much scrutiny as the hockey stick graph. Critics attacked everything from the statistical methods to the proxy selection to the blending of instrumental and proxy data. They suggested the distinctive shape might be an artifact of methodology rather than a real climate signal (RealClimate, 2021).

However, the most important scientific test is replication, and that's where the hockey stick truly demonstrated its resilience. In the decades since the original publication, numerous independent teams using different statistical methods, different proxy combinations, and different approaches have produced reconstructions remarkably similar to the original hockey stick. It's like when multiple people independently come up with your exact birthday as their ATM PIN—it's no longer a coincidence but a significant pattern (IPCC, 2013).

Today, the scientific consensus supports the hockey stick's main conclusions, though with greater appreciation for the uncertainties and regional variations in past climate. The original hockey stick has been joined by a whole hockey team of similar reconstructions, all pointing to the same unsettling conclusion: current warming is unprecedented in both rate and magnitude compared to natural variations of the past millennium (IPCC, 2013).

The Divergence Problem: When Trees Get Confused

Nothing in science is ever perfectly straightforward, and tree-ring climate reconstructions are no exception. One of the most intriguing challenges in dendroclimatology is the "divergence problem"—the observation that some tree-ring width and density measurements at certain high-latitude sites show a weakening relationship with temperature since roughly the mid-20th century (ScienceDirect, 2013).

In simple terms, tree rings at some sites stopped tracking temperature as reliably as they had for centuries. While temperatures rose, tree-ring widths or densities didn't increase as the historical relationship would predict. It's like your previously reliable friend who always told you honestly if your outfit looked good suddenly becoming suspiciously complimentary about everything you wear (ScienceDirect, 2013).

Possible Explanations

Scientists have proposed several potential explanations for this divergence:

1. **Drought stress**: Higher temperatures without corresponding increases in precipitation could create moisture stress that limits growth despite warmth.
2. **Pollution effects**: Air pollution might be affecting tree growth in ways that mask the temperature signal.
3. **Global dimming**: Anthropogenic aerosols reduced sunlight reaching Earth's surface during parts of the 20th century, potentially affecting tree growth independently of temperature.
4. **Non-linear responses**: Trees might have threshold responses where the temperature-growth relationship changes beyond certain temperature ranges.
5. **Methodological issues**: Some of the divergence might relate to how the data is processed and standardized rather than actual tree behavior.

Importantly, the divergence problem doesn't invalidate tree-ring reconstructions—it's observed only in some high-latitude sites, affects the recent period for which we have instrumental records anyway, and doesn't appear in many proxy records. It's more like a footnote to the climate story than a plot hole that undermines the entire narrative (IPCC, 2013).

However, this scientific puzzle does highlight the importance of using multiple proxy types and carefully validating tree-ring chronologies against instrumental data during the overlap period. It's a reminder that all proxy records have limitations and that the strongest climate reconstructions combine multiple lines of evidence rather than relying on a single proxy type. It's like how you wouldn't hire someone based solely on their résumé without also checking references, conducting an interview, and stalking them on social media like a normal hiring manager (IPCC, 2013).

Beyond Trees: The Multi-Proxy Approach

While tree rings provide incredible insights into past climate, they have limitations. They're geographically restricted to where trees grow, typically reflect growing season conditions rather than annual averages, and only extend back thousands (not millions) of years. To build a more complete climate history, scientists combine tree rings with other "proxy" records—natural archives that indirectly record climate information (NASA, 2017).

This multi-proxy approach is like assembling a team of witnesses with different perspectives on the same event—each has limitations, but together they create a more complete picture:

Ice Cores: Frozen Time Capsules

Ice cores drilled from ice sheets and glaciers contain annual layers with air bubbles that preserve ancient atmosphere samples, dust particles, and chemical isotopes that reflect temperature. The oldest continuous ice cores from Antarctica extend back about 800,000 years—much further than tree rings but still a blink in geological time (IPCC, 2013).

What makes ice cores uniquely valuable is that they contain actual samples of ancient air, allowing direct measurement of past greenhouse gas concentrations. The verdict? Current atmospheric CO_2 levels (above 410 ppm) far exceed anything seen in the past 800,000 years when concentrations oscillated between about 180-280 ppm. That's like discovering that what you thought was a moderately high credit card balance is actually higher than anything your family has carried for the past 30 generations (IPCC, 2013).

Corals: Ocean Climate Archives

Like trees, corals build annual growth bands that reflect ocean conditions, including temperature, salinity, and nutrient availability. By analyzing the chemical composition and physical characteristics of these bands, scientists can reconstruct sea surface temperatures, rainfall patterns, and even ocean circulation changes (IPCC, 2013).

Corals provide critical information about tropical ocean regions where tree records are sparse. While most coral records only extend back several centuries, some exceptionally old coral colonies and fossil corals provide windows into ocean conditions thousands of years ago. Unfortunately, these valuable climate recorders are now themselves threatened by the climate changes they're helping us understand—like a historian documenting their own demise (NASA, 2017).

Lake and Ocean Sediments: Mud with a Message

Lake and ocean sediments accumulate in layers that contain pollen, microfossils, chemical signatures, and physical properties reflecting environmental conditions at the time of deposition. These sediment records can extend back millions of years, providing climate information well beyond the reach of tree rings or ice cores (IPCC, 2013).

Sediment cores have revealed dramatic climate shifts throughout Earth's history, including rapid warming events like the Paleocene-Eocene Thermal Maximum about 56 million years ago, when a massive release of carbon into the atmosphere caused global warming of 5-8°C over several thousand years. Even this ancient "natural" climate disruption occurred much more slowly than current human-caused warming—it's like comparing your grandparent's cautious driving to your teenager doing donuts in the parking lot (IPCC, 2013).

Cave Formations: Drip by Climate Drip

Stalagmites and stalactites (speleothems) forming in caves preserve detailed records of precipitation, temperature, and vegetation changes in their growth layers and chemical composition. Water dripping through soil and rock dissolves minerals that later precipitate in caves, creating these formations with chemical signatures that reflect surface conditions (IPCC, 2013).

Some cave formations can be precisely dated using radiometric techniques and provide continuous climate records spanning hundreds of thousands of years. They're particularly valuable for reconstructing monsoon patterns and regional precipitation changes in areas where other proxy records might be limited. Think of them as nature's rainfall logbooks, meticulously updated drop by drop over millennia (IPCC, 2013).

Bringing It All Together: What the Full Proxy Record Tells Us

When all these climate proxies—tree rings, ice cores, corals, sediments, cave formations, and others—are analyzed together, they tell a consistent and concerning story about Earth's climate:

1. Current Warming Is Unprecedented

Multiple independent proxy reconstructions confirm that current global temperatures are higher than at any time in at least the past 2,000 years, and the rate of warming exceeds anything detected in much longer records. It's not just that it's warm—it's the speed of change that's truly exceptional. It's like comparing the gradual weight gain of middle age to suddenly putting on 20 pounds in a month—both involve gaining weight, but one suggests something seriously wrong is happening (IPCC, 2013).

2. Current CO_2 Levels Are Off the Charts

Ice core data unequivocally shows that current atmospheric CO_2 concentrations exceed anything experienced on Earth in at least 800,000 years, and geological evidence suggests we may have to go back millions of years to find comparable levels. This extra greenhouse gas loading is the smoking gun connecting human activities to observed warming. It's like finding your fingerprints all over the empty cookie jar—sure, there could theoretically be another explanation, but let's be real about who ate all the cookies (IPCC, 2013).

3. Climate Can Change Abruptly

Proxy records reveal that while climate often changes gradually, it can also shift remarkably quickly under certain conditions. Ice cores from Greenland show that regional temperatures sometimes changed by 10°C or more within decades during the last ice age—a sobering reminder that the climate system contains tipping points and feedback mechanisms that can accelerate change once certain thresholds are crossed. It's like how your teenager's room can maintain a relatively stable level of messiness for months before suddenly crossing a threshold into biohazard territory over a single weekend (IPCC, 2013).

4. Natural Factors Can't Explain Recent Warming

The proxy record clearly identifies the major natural factors that influenced climate before human interference—solar variations, volcanic eruptions, orbital changes, and internal climate system variability. When scientists include all these natural factors in climate models, they cannot reproduce the observed warming without adding the effect of human greenhouse gas emissions. It's like trying to explain why your car isn't starting without mentioning that it's out of gas—you can talk about battery age, starter motors, and spark plugs all day, but you're missing the obvious primary cause (IPCC, 2013).

From Past to Future: What History Teaches Us About What's Coming

The detailed climate history revealed by tree rings and other proxies doesn't just tell us where we've been—it provides crucial context for understanding where we're heading:

Learning from Past Warm Periods

Proxy records of past warm periods like the mid-Pliocene (about 3 million years ago, with CO_2 levels similar to today) provide glimpses of what a warmer Earth might look like. During this period, global temperatures were 2-3°C warmer than pre-industrial levels, and sea levels were 15-25 meters higher due to reduced ice sheets in Greenland and Antarctica (IPCC, 2013).

These ancient analogs aren't perfect predictors—today's warming is occurring much faster, and modern geography and ecosystems differ from those of the past. But they do illustrate the potential magnitude of changes once the climate system fully adjusts to current greenhouse gas levels. It's like how your friend who gained 30 pounds after college gives you a preview of your possible future if you continue your current pizza-and-Netflix lifestyle—not guaranteed, but certainly worth considering as you reach for another slice (IPCC, 2013).

Tipping Points and Nonlinear Responses

Perhaps the most concerning lesson from paleoclimate records is that climate change doesn't always proceed in a smooth, linear fashion. Proxy evidence reveals numerous instances of abrupt shifts when the climate system crossed critical thresholds or tipping points (IPCC, 2013).

For example, sediment cores from the North Atlantic show that ocean circulation patterns have reorganized rapidly in the past, with dramatic effects on regional climates. Archaeological and tree ring evidence reveals civilizations that collapsed after crossing environmental thresholds, such as the Maya, who suffered as changing rainfall patterns undermined their agricultural system during an extended drought. It's like how you can charge "just a little more" on your credit card each month until suddenly you cross a threshold where minimum payments exceed your budget and the whole financial system collapses (IPCC, 2013).

Understanding these nonlinear responses is crucial because they mean climate impacts might not emerge gradually but could appear suddenly once certain thresholds are crossed. It's the climatic equivalent of the old saying about bankruptcy—it happens gradually, then suddenly.

The "Business as Usual" Scenario Looks Scary

When scientists compare current emission trajectories with proxy records of past climates, the implications are sobering. Without substantial emissions reductions, we're on track for warming of 3-4°C by 2100—conditions Earth hasn't experienced in millions of years, during times when our ancestors were still developing opposable thumbs rather than TikTok dances (IPCC, 2013).

The proxy record suggests such warming would eventually lead to:

- Sea level rise of many meters (though taking centuries to fully manifest)
- Dramatic shifts in precipitation patterns and ecosystem boundaries
- Expansion of uninhabitable zones due to excessive heat and humidity
- Mass extinctions as adaptation and migration capacities are exceeded

These aren't wild speculations but logical extensions of what proxy records tell us about Earth's sensitivity to greenhouse gas concentrations. It's like how examining the relationship between calorie intake and weight gain over the past year can reasonably predict where your waistline is headed if you continue your current donut consumption trajectory (IPCC, 2013).

The Good News: Climate Response to Emission Reductions

While proxy records highlight risks, they also offer hope by confirming that the climate system responds to changes in atmospheric composition. Just as increasing greenhouse gases warm the planet, reducing them can eventually stabilize the climate—though with lag times due to ocean thermal inertia and other factors (IPCC, 2013).

The relationship between CO_2 and temperature in the proxy record helps constrain estimates of "climate sensitivity"—how much warming results from doubling atmospheric CO_2. These estimates suggest that rapid emission reductions that limit CO_2 levels to around 450 ppm could potentially keep warming below 2°C relative to pre-industrial levels—the threshold identified as avoiding the most dangerous climate impacts (IPCC, 2013).

The proxy record thus provides both warning and guidance—showing the enormous risks of continued high emissions while confirming that alternative pathways remain possible if we act quickly enough. It's like a medical test that both diagnoses a serious condition and confirms it's treatable if addressed promptly. The test results are concerning, but at least treatment options exist (IPCC, 2013).

When Scientists Become Tree Huggers (Literally)

The science of reading climate history from trees might seem abstract, but it involves surprisingly hands-on fieldwork. Dendroclimatologists don't just analyze samples in sterile laboratories—they trek through forests, climb mountains, and sometimes even hug trees (though for sampling purposes, not emotional support) (NOAA, 2018).

Field Adventures in Climate Science

Collecting tree-ring samples typically involves using a specialized tool called an increment borer—essentially a hollow drill bit that extracts a pencil-sized core from the tree without harming it (much like a biopsy). The process requires careful site selection, physical exertion, and occasionally creative problem-solving when equipment fails or weather turns nasty (NASA, 2017).

Some memorable dendroclimatology fieldwork has included:

- Sampling 5,000-year-old bristlecone pines at elevations above 10,000 feet in the White Mountains of California
- Collecting subfossil logs from river cutbanks in Siberia while fending off mosquitoes the size of small drones
- Diving to retrieve ancient wood preserved in lake bottoms (combining dendrochronology with SCUBA certification)
- Analyzing beams from historic buildings and archaeological sites to extend chronologies back in time

It's science that combines technical precision with adventure—like Indiana Jones with more graph paper and fewer Nazi encounters (Nature, 2022).

The Human Stories Behind the Science

The history of dendroclimatology includes some remarkable human stories. A.E. Douglass, the field's founder, initially searched for connections between tree rings and sunspot cycles to support his astronomical research. While that particular hypothesis didn't pan out as expected, it led him to discover something even more valuable—a method for precisely dating archaeological ruins in the American Southwest and reconstructing past climate conditions (Carbon Brief, 2018).

Similarly, Edmund Schulman spent years searching for the world's oldest trees, finally discovering ancient bristlecone pines in California's White Mountains in the 1950s. His work extended verifiable tree-ring chronologies back thousands of years and laid the groundwork for radiocarbon dating calibration. Sadly, Schulman died of a heart attack at age 49 in 1958, shortly after his most significant discoveries—a reminder that while trees often live for millennia, the scientists who study them aren't always so fortunate (NOAA, 2018).

These personal dimensions remind us that climate science isn't conducted by faceless institutions or autonomous algorithms but by real people with curiosity, dedication, and occasionally questionable field hygiene after weeks in remote locations.

The Most Important Tree Ring: The One Forming Right Now

As we've explored the fascinating science of tree rings and other climate proxies, it's worth considering a sobering thought: future scientists will study the tree rings forming right now for evidence of our climate impact. Trees growing today are documenting our choices, and their growth rings will testify to future generations about what we did—or failed to do—about climate change (IPCC, 2013).

What Will Our Chapter in Tree-Ring History Show?

Will the tree rings of the mid-21st century show continued warming, more frequent drought stress, and increasingly erratic growth patterns as climate disruption accelerates? Or will they reveal a stabilization period as human emissions decline and the climate system begins to rebalance? Either way, the evidence will be written in wood, waiting for future scientists to analyze (IPCC, 2013).

In a very literal sense, we are creating climate history that will be studied centuries from now. Future dendroclimatologists will extract cores from trees alive today and use them to reconstruct our climate choices. Those researchers might wonder what we were thinking as we watched the evidence accumulate but hesitated to take decisive action. It's like how we look back at lead paint or leaded gasoline and think, "How could they not have realized the harm they were causing?"—except we don't have the excuse of ignorance (USGS, 2022).

Your Life in Tree Rings

On a more personal note, think about this: if you live near trees, they're recording the climate conditions of your lifetime in their growth rings. Trees planted when you were born have been creating a living record of every year you've experienced. The wet years, the drought years, the heat waves, the cold snaps—all documented in concentric circles of wood (NOAA, 2018).

There's something profound about this parallel biological record keeping. Long after your social media posts have been deleted and your digital photos have corrupted, trees might still harbor the physical evidence of what the climate was like during your lifetime. It's a reminder of our temporary presence in a much longer biological and climatic story—and our outsized responsibility at this crucial moment in that story (IPCC, 2013).

Conclusion: The Rings Don't Lie

As we close this arboreal adventure through climate history, what can we conclude from all these tree rings, ice layers, and sediment cores? The evidence is clear and consistent:

1. **Earth's climate has changed naturally in the past**, sometimes dramatically, in response to various forcing factors like orbital variations, solar changes, and volcanic activity.

2. **Current changes are different** in critical ways—they're occurring much faster, they're driven primarily by human greenhouse gas emissions, and they're happening while human civilization is dependent on the relatively stable climate of the past several thousand years.

3. **The past provides both warnings and guidance**—showing the risks of major climate disruption while also helping us understand how the climate system responds to different forcing factors.

4. **We're conducting a planetary experiment** without fully knowing the consequences—pushing the climate system into a state not experienced for millions of years, at a rate unprecedented in geological history.

The science of paleoclimatology doesn't just satisfy academic curiosity about the past—it provides essential context for understanding our present predicament and future options. Without this long-term perspective, we might mistakenly view current changes as minor fluctuations rather than the significant departures they represent (IPCC, 2013).

As we move forward, these ancient climate archives remind us of something essential: Earth's climate can change dramatically, human activities are now the dominant force driving that change, and our choices today will be recorded in tree rings, ice layers, and sediment cores for future scientists to study. The trees are watching, and they're keeping meticulous records. Let's give future dendrochronologists something hopeful to analyze, shall we?

In the next chapter, we'll examine how money flows through the climate system—from fossil fuel financing to climate solution investments—and how redirecting these financial currents may be one of our most powerful tools for addressing the climate crisis. As it turns out, following the money reveals as much about our climate situation as following the science.

Chapter 9: Banks, Gas Tanks, and Solar Panels

Follow the Money in Climate Change

Remember when you were a kid and your piggy bank was your most prized financial institution? Those were simpler times. You'd drop in a quarter, shake it around, and dream about all the candy you could buy. These days, grown-up banks are shaking things up too—except instead of candy, they're deciding the fate of our planet's climate. No pressure or anything.

Banks might seem like boring places where your money sits around in digital vaults, occasionally venturing out to pay for your coffee addiction. But in reality, these financial powerhouses are secretly climate supervillains AND potential climate superheroes. It's a confusing duality, like finding out your friendly neighborhood accountant is also a part-time ninja.

Banks are a major source of both the problem and the solution when it comes to climate change (Rainforest Action Network, 2024). The financial decisions made in sleek boardrooms have consequences that ripple across our warming planet—from the Arctic ice sheets to your increasingly sweaty summer electric bill.

This chapter follows the money trail in climate change—from fossil fuel financing to renewable energy investments—and explores how our economic decisions are either fueling the climate crisis or paving the way to a sustainable future. So grab your calculator and maybe a stress ball. We're about to make climate finance almost as interesting as your last Netflix binge.

The Banking Climate Contradiction

Here's a brain teaser for you: What's the connection between your checking account and a melting glacier? The answer: more than you might think.

Banks are essentially massive pools of money looking for places to multiply. They take your deposits and make loans and investments across the economy. And for decades, they've had a love affair with fossil fuels that would make a romance novelist blush. Why? Because historically, oil, gas, and coal have been reliable money-makers.

According to recent research, the world's 60 largest banks have provided a staggering $6.9 trillion in financing to fossil fuel companies since the 2015 Paris Agreement (Rainforest Action Network, 2024). That's "trillion" with a "t"—or roughly the GDP of Australia being poured into the very industries driving climate change. It's like paying your arsonist neighbor to buy more matches while your house is already on fire.

This massive financial support creates a major obstacle to addressing climate change, as these investments lock in carbon emissions for decades to come (Banking on Climate Chaos, 2024). When a bank finances a new oil pipeline or coal plant, they're essentially betting against our clean energy future.

But here's where things get interesting. The same banks funding fossil fuels are simultaneously ramping up their green financing. It's financial climate schizophrenia—like ordering both a salad and a triple bacon cheeseburger and convincing yourself it all balances out.

JPMorgan Chase, for example, has been one of the largest fossil fuel financiers globally, yet they've also committed billions to clean energy projects and sustainability initiatives (Sierra Club, 2024). It's the financial equivalent of installing solar panels on your Hummer.

This contradiction exists because banks are caught in a climate transition. They see the writing on the wall (or rather, the rising temperatures on the thermometer), but they're still addicted to those reliable fossil fuel returns. It's the awkward teenage phase of climate finance—complete with mood swings and identity crises.

Carbon Bubbles and Stranded Assets

Remember the housing bubble that led to the 2008 financial crisis? Well, climate scientists and economists warn that we could be facing a "carbon bubble" that makes that look like a tiny hiccup.

The carbon bubble represents the overvaluation of fossil fuel reserves that will become worthless if the world takes serious action on climate change. Here's the uncomfortable math: to keep global warming below 2°C (as agreed in the Paris Agreement), we can only burn about 20-30% of existing fossil fuel reserves. The rest must stay in the ground—permanently (IPCC, 2022).

This creates a concept economists call "stranded assets"—fossil fuel reserves that will never be extracted because doing so would cook the planet. When these assets are eventually recognized as worthless, the financial impact could be enormous.

What's particularly concerning is that many pension funds, insurance companies, and yes, your own investments, may be exposed to this carbon bubble. It's like discovering your retirement savings are tied up in Blockbuster Video stock circa 2010. Not ideal.

The financial implications go beyond fossil fuel companies. Banks that have loaned billions to coal mines, oil drillers, and pipeline companies could face substantial losses if these assets become stranded. The ripple effects could touch everything from real estate values in coastal areas to insurance rates in wildfire-prone regions.

Some forward-thinking financial institutions are already stress-testing their portfolios against climate scenarios. Others are burying their heads deeper in the (oil) sand, perhaps hoping climate change will magically reverse itself if ignored hard enough. (Spoiler alert: it won't.)

Follow the Green: Renewable Energy Investment

If you're feeling depressed after that last section, here's some good news: renewable energy investment is absolutely booming. It's like watching the awkward sustainable energy kid from high school become the coolest trend-setter a decade later.

Global investment in renewable energy has grown exponentially, reaching $1.8 trillion in 2023 for all low-carbon energy transition technologies (BloombergNEF, 2024). This isn't just good for the planet —it's increasingly good business. The cost of solar panels has dropped by about 89% since 2010, while wind turbine costs have fallen by approximately 70% during the same period (IRENA, 2023).

Renewable energy is coming of age at a critical moment in our climate journey. Massive offshore wind projects like Vineyard Wind off the coast of Massachusetts exemplify the scale that renewables are now achieving (U.S. Department of Energy, 2023).

Banks and investment firms are following this trend, dramatically increasing their clean energy portfolios. Investment giants like BlackRock have made climate change central to their investment strategy, while dedicated green banks and climate funds are emerging worldwide (World Economic Forum, 2023).

Even traditional energy companies are pivoting toward renewables. Many oil and gas majors are rebranding themselves as "energy companies" rather than "oil companies"—a subtle but significant shift that acknowledges the changing landscape. It's like when your favorite junk food brand suddenly introduces a "healthier option"—except this time, the health of the entire planet depends on it.

The financial case for renewables continues to strengthen as technology improves, costs fall, and climate policies evolve. Investment in clean energy isn't just an ethical choice anymore—it's increasingly the smart money move. Green is the new black, both environmentally and financially.

Carbon Markets and Pricing: Making Pollution Pay

Imagine if every time you farted in an elevator, you had to pay a small fee. You'd probably think twice before eating that bean burrito for lunch, right? That's essentially the idea behind carbon pricing—making polluters pay for their emissions.

Carbon markets and pricing mechanisms attempt to incorporate the true cost of carbon emissions into the economy. They come in various forms:

1. **Carbon taxes**: Direct fees imposed on the carbon content of fuels

2. **Cap-and-trade systems**: Markets where companies buy and sell permitted carbon allowances
3. **Internal carbon pricing**: When companies voluntarily apply a carbon price to their decision-making

These approaches aim to shift investment away from high-carbon activities toward cleaner alternatives. It's like putting your thumb on the economic scale—but in this case, it's to correct a massive market failure that threatens civilization as we know it. So, you know, no big deal.

Carbon markets have had a rocky history. The European Union's Emissions Trading System (EU ETS), the world's largest carbon market, initially suffered from an oversupply of permits that kept prices too low to drive meaningful change. It was like trying to discourage speeding by issuing $1 tickets—more of a minor inconvenience than a real deterrent.

However, reforms have strengthened many carbon pricing systems, and the idea is gaining global momentum. In 2023, carbon pricing revenues reached a record $104 billion globally, with 75 carbon pricing instruments in operation worldwide (World Bank, 2024). Even in the United States, which has historically resisted national carbon pricing, regional initiatives like the Northeast's Regional Greenhouse Gas Initiative (RGGI) have proven effective.

The financial sector plays a crucial role in carbon markets—developing trading platforms, creating carbon-linked financial products, and helping companies manage their carbon liabilities. Some banks have even developed "green bonds" and other instruments specifically designed to fund low-carbon projects.

For everyday investors, carbon markets create new opportunities to align investments with climate goals. Climate-aware investment products have exploded in popularity, from low-carbon index funds to green bonds and direct investment in carbon offset projects.

Carbon pricing represents one of the most powerful tools for redirecting capital toward a low-carbon future. By putting a price on pollution, we can harness the power of markets to solve the very problem they helped create. It's fighting fire with fire—or in this case, fighting climate change with capitalism.

Divestment: Voting with Your Dollars

Remember when you threatened to take your toys and go home if your childhood friend didn't play by the rules? The fossil fuel divestment movement works on a similar principle—except the toys are trillions of dollars in investments, and the "rules" are the physical limits of our planet.

Divestment means removing investments from fossil fuel companies as a form of economic and moral pressure. It started on college campuses around 2011 but has since grown into a global movement. Major institutions including universities, religious organizations, pension funds, and even some sovereign wealth funds have committed to divesting from fossil fuels.

The financial impact of divestment goes beyond the direct removal of capital. It creates stigmatization around fossil fuel investments, shifts public discourse, and can eventually affect companies' ability to secure financing. It's like the financial equivalent of being uninvited to the cool kids' party.

Critics argue that divestment is merely symbolic—that sold shares are simply purchased by less climate-conscious investors. But proponents counter that the movement's power lies in changing social norms around fossil fuel investments and building political support for climate policies (Ansar et al., 2023).

The divestment movement has achieved some remarkable victories. In 2021, Harvard University announced it would divest its $42 billion endowment from fossil fuels after years of student activism. When an institution that old and wealthy changes course, it's like watching your stubborn grandfather finally admit that maybe, just maybe, smartphones aren't just a passing fad.

For individual investors, divesting personal portfolios from fossil fuels has become increasingly accessible through sustainable investment options. Many financial advisors now offer fossil-free investment strategies, and major investment firms provide climate-conscious funds.

This approach allows ordinary people to align their savings with their values—ensuring their money isn't undermining the future they're saving for (350.org, 2023). It's personal finance meets climate activism: your retirement account can now be part of the solution rather than part of the problem.

The divestment movement represents a democratic approach to climate finance—demonstrating that even in a world dominated by big banks and multinational corporations, collective action by many smaller investors can drive significant change. It turns out your piggy bank might have more power than you thought.

Greenwashing: When Banks Wear Eco-Costumes

Picture this: A bank runs heartwarming commercials showing wind turbines and smiling children, publishes sustainability reports filled with verdant imagery, and makes vague promises about a "greener future"—all while pumping billions into new oil pipelines and coal plants. Welcome to the world of greenwashing.

Greenwashing happens when financial institutions exaggerate or misrepresent their environmental commitments. It's like putting a "low fat" label on a chocolate cake—technically, it might contain less fat than some other cake, but it's still not exactly health food.

Companies use environmental marketing to appear more eco-friendly than they actually are, potentially misleading consumers and investors (Delmas and Burbano, 2023).

Common greenwashing tactics in the financial sector include:

1. **Setting distant net-zero targets** without clear interim goals or implementation plans
2. **Cherry-picking statistics** that highlight small green initiatives while ignoring much larger fossil fuel activities
3. **Using vague terminology** like "sustainable," "green," or "eco-friendly" without specific definitions
4. **Creating green-labeled financial products** with minimal environmental standards

Detecting greenwashing requires looking beyond glossy sustainability reports and press releases. Investors need to examine concrete metrics like:

- The bank's fossil fuel financing relative to renewable energy investments
- Whether climate commitments have specific timelines and accountability mechanisms
- If the institution supports or opposes climate policies
- How executive compensation connects to climate performance

Regulatory bodies are increasingly cracking down on greenwashing. The European Union's Sustainable Finance Disclosure Regulation (SFDR) and taxonomy requirements aim to create standardized definitions of "green" investments and force greater transparency (European Commission, 2023).

For individuals, avoiding greenwashed financial products means doing homework beyond marketing materials. Tools like fossil fuel financing report cards and sustainable investment certifications can help separate genuine climate leaders from green posers.

The best defense against greenwashing is knowledge. Understanding the difference between meaningful climate action and eco-marketing fluff allows consumers and investors to reward genuine sustainability and punish environmental pretenders. It's applying the old adage "trust, but verify" to climate finance—except maybe with a bit less trust and a lot more verification.

Climate Risk: Coming to a Balance Sheet Near You

Climate change isn't just an environmental crisis—it's increasingly seen as a financial risk that threatens economic stability. This realization is transforming how banks, insurers, and regulators approach climate issues.

Climate risks come in two main flavors:

1. **Physical risks** from extreme weather, sea-level rise, and other direct climate impacts
2. **Transition risks** from policy changes, technology shifts, and market preferences during the shift to a low-carbon economy

These risks are already materializing. Hurricanes, floods, and wildfires have caused billions in damages and bankruptcy for major companies like Pacific Gas & Electric (NOAA, 2024). Meanwhile, rapid shifts in energy markets have left some fossil fuel assets stranded sooner than expected.

Financial regulators worldwide are responding by incorporating climate into their supervision frameworks. The Network for Greening the Financial System (NGFS), a coalition of central banks and supervisors, considers climate change "a source of financial risk" that falls within their mandates to ensure financial stability (NGFS, 2023).

Major financial institutions are beginning to conduct climate stress tests —simulations that assess how portfolios would perform under various climate scenarios. These tests examine both physical risks (like mortgage exposure in flood-prone areas) and transition risks (like loans to carbon-intensive industries).

Disclosure is another key trend. Initiatives like the Task Force on Climate-related Financial Disclosures (TCFD) have developed frameworks for companies to report their climate risks and opportunities. This transparency helps investors make informed decisions and encourages companies to better manage their climate exposures (TCFD, 2023).

For individual investors, climate risk awareness means reconsidering conventional wisdom about "safe" investments. Coastal real estate, infrastructure vulnerable to extreme weather, and carbon-intensive industries all carry new forms of risk in a changing climate.

The integration of climate into financial risk management represents a profound shift in how markets view environmental issues—from purely ethical concerns to material financial considerations. When conservative risk managers at central banks start worrying about climate change, you know the conversation has moved beyond tree-huggers and environmental activists.

As one finance professional put it: "The climate doesn't care about your investment thesis." Financial systems are finally catching up to this reality, recognizing that no portfolio is immune to a destabilized climate. In this sense, financial self-interest and environmental protection are increasingly aligned—perhaps our best hope for driving the massive capital reallocation needed to address climate change.

The Human Side: Financial Inclusion in the Climate Transition

Climate finance isn't just about billion-dollar deals and complex derivatives. It's also about ensuring that regular people—especially those in vulnerable communities—can participate in and benefit from the green economy. Because a clean energy future isn't much good if only the wealthy get to enjoy it.

The concept of a "just transition" recognizes that shifting from fossil fuels to clean energy affects communities differently. Coal miners, oil workers, and frontline communities often bear disproportionate costs during economic transformations. Just climate finance aims to support these communities while expanding clean energy access to those who need it most.

Community solar programs offer one promising model. These shared solar facilities allow multiple participants to benefit from a single solar installation, making clean energy accessible to renters, lower-income households, and those with unsuitable roofs. Such initiatives are part of the solution to democratizing renewable energy (U.S. Department of Energy, 2024).

Green banks and climate funds increasingly incorporate equity goals into their missions. This approach ensures climate finance reaches beyond affluent early adopters.

Microfinance for climate resilience represents another innovative approach. Small loans for household solar systems, efficient appliances, or climate-resilient agriculture help vulnerable communities adapt to changing conditions while reducing emissions. These programs recognize that climate solutions look different for a subsistence farmer in Bangladesh than for a suburban homeowner in Boston.

Financial inclusion also means addressing the urban-rural divide in climate finance. While coastal cities often attract significant green investment, rural communities frequently lack access to climate capital despite their crucial role in land management, agriculture, and distributed energy generation.

For the average person, participating in climate finance doesn't require millions in investment capital. Options include:

- Opening accounts at community development financial institutions (CDFIs) that support local sustainability

- Joining credit unions with strong environmental commitments
- Investing in climate-focused mutual funds with low minimum requirements
- Supporting crowdfunded renewable energy projects

The democratization of climate finance represents one of the most important developments in the field. By expanding who participates in and benefits from the green economy, we can build broader political support for climate action while addressing historical inequities. Because saving the planet shouldn't be an exclusive club for the already privileged.

Your Money, Your Climate: Personal Finance in a Warming World

Let's bring this home: What does all this mean for your personal finances? Whether you have millions invested or are just scraping by with a modest savings account, your financial decisions intersect with climate change in surprising ways.

Start with your banking relationship. Where you keep your checking and savings accounts matters. Major banks use deposits to fund their lending activities—including fossil fuel financing. Credit unions and smaller community banks often have substantially smaller carbon footprints, while dedicated ethical banks explicitly avoid fossil fuel investments.

For those with investment portfolios, climate considerations have never been more relevant. Beyond ethical concerns, climate risks now present material financial considerations. Options for climate-conscious investing have expanded dramatically, from broad ESG (Environmental, Social, Governance) funds to targeted investments in renewable energy, electric vehicles, and climate adaptation.

Personal financial choices contribute to broader systemic change (BloombergNEF, 2024). When millions of people shift their money in similar directions, it creates powerful market signals.

Insurance represents another financial frontier affected by climate change. As extreme weather events become more frequent, insurance premiums in vulnerable areas are rising, and coverage is becoming harder to secure. Smart homebuyers now consider climate risks alongside traditional factors like school districts and commute times.

Retirement planning also demands climate awareness. Someone in their 30s today will retire into a world shaped by how we address—or fail to address—climate change in the coming decades. This long-term perspective aligns naturally with sustainable investing approaches focused on systemic risks and opportunities.

Even everyday purchasing decisions connect to climate finance. Credit cards from environmentally committed institutions or those offering carbon offsets allow consumers to align daily transactions with climate values. Some apps even round up purchases and invest the difference in climate solutions.

The key insight is that personal finance and climate action aren't separate domains—they're increasingly interconnected. By bringing climate consciousness to our financial lives, we can build resilience against climate risks while contributing to solutions.

These individual actions may seem small, but collectively they create the economic and political momentum needed for systemic change (Project Drawdown, 2023). Your money is constantly voting for the future you want to live in. Make sure it's voting for a livable climate.

Conclusion: Financial Gravity and Climate Solutions

Money, like water, flows along the path of least resistance. For too long, our financial systems have channeled capital toward fossil fuels and carbon-intensive industries—not because bankers hate polar bears, but because that's where the economic incentives pointed.

Addressing climate change requires redirecting these massive capital flows toward clean energy, sustainable infrastructure, and climate resilience. It's less about inventing new technologies (though that helps) and more about financing the massive deployment of solutions we already have.

The good news is that financial gravity is beginning to shift. Renewable energy is increasingly outcompeting fossil fuels on cost alone. Climate risks are being priced into markets. Consumers and investors are demanding climate accountability. And policymakers are (slowly) implementing frameworks that make carbon pollution less profitable and clean solutions more attractive.

This financial transformation won't happen overnight. Entrenched interests resist change, and economic systems have considerable inertia (Mercure et al., 2023). But the direction of travel is increasingly clear. The question isn't whether finance will align with climate goals, but how quickly and through what combination of market forces, policy interventions, and social pressure.

For individuals, this shifting landscape presents both challenges and opportunities. Climate-aware financial decisions can protect personal assets from emerging risks while contributing to broader solutions. Whether you're a billionaire investor or just someone with a checking account, your financial choices matter in the climate equation.

The relationship between money and climate change works both ways. Finance has fueled the climate crisis, but it also holds the keys to addressing it. By redirecting capital from the problem to the solution, we can turn our financial system from climate villain into climate hero.

Perhaps the most important insight is that financial considerations and climate progress aren't opposing forces—they're increasingly aligned. Clean energy isn't just good for the planet; it's becoming the smart money move. This alignment of profit and planetary health may be our best hope for the rapid, massive economic transformation we urgently need.

In the end, climate finance isn't just about money—it's about values, choices, and the future we want to create. Because when it comes to addressing climate change, following the money tells us everything about where we've been and where we're headed.

In our next chapter, we'll examine the electricity infrastructure that powers our modern world and the challenges of transitioning it to clean energy sources. The grid that delivers power to your home might not seem sexy, but it's about to get one of the biggest makeovers in history —and we're all going to be affected.

Chapter 10: The Grid Isn't Ready, and Neither Are We

Shockingly Unprepared

Remember the last time your power went out? That brief interruption probably launched you into a panic spiral about your phone battery, food spoilage, and whether you still remember how to entertain yourself without streaming services. Now multiply that helpless dependency by an entire civilization's electricity addiction, and you've got the central challenge of the clean energy transition.

Our electrical grid—that complex web of wires, transformers, and infrastructure delivering power to your home—was designed for a fossil fuel world that we urgently need to leave behind. It's essentially the world's largest machine, an interconnected nervous system spanning continents, powering everything from hospital ventilators to your cousin's embarrassing TikTok dances. And right now, it's being asked to undergo radical heart surgery while still pumping blood to every extremity. No pressure.

This chapter explores the monumental challenge of modernizing our electricity infrastructure for the clean energy revolution. It's like trying to replace every water pipe in your house while still taking showers, washing dishes, and keeping your houseplants alive. We're talking about one of the most difficult engineering projects in human history, and it needs to happen faster than your toddler can find the electrical outlet you forgot to childproof.

Grid Basics: What Even Is This Thing?

Before diving into why the grid is so spectacularly unprepared for our clean energy ambitions, let's understand what we're working with. The electric grid is essentially a three-part system: generation (power plants creating electricity), transmission (high-voltage lines transporting electricity over long distances), and distribution (lower-voltage local networks delivering electricity to end users).

In the traditional grid, electricity flows in one direction: from large centralized power plants (usually burning coal, natural gas, or uranium) through transmission lines to distribution networks and finally to your home. It's like a river with tributary streams—everything flows from big to small, from centralized production to distributed consumption (Department of Energy [DOE], 2022).

This system was designed with a simple assumption: electricity demand is unpredictable, but supply can be controlled. When you flip a light switch, a power plant somewhere adjusts its output to match. It's like having a water utility that somehow knows exactly when you're going to flush your toilet and immediately pumps the exact right amount of water through the system. Pretty impressive, when you think about it.

The traditional grid relies on "dispatchable" power sources—plants that can be turned up or down as needed. Need more electricity during a heat wave when everyone's cranking their air conditioning? Just burn more coal or natural gas. This controllable supply was the perfect match for our unpredictable energy demands—at least until we realized that burning fossil fuels is cooking the planet faster than your uncle overcooks burgers on the Fourth of July.

Renewables: Plot Twist for the Grid

Renewable energy sources like wind and solar introduce a fundamental change to this system: they're "non-dispatchable," meaning we can't control when they generate electricity. Unless you've invented some kind of weather-controlling device (in which case, please put down this book and call the nearest university immediately), we can't make the sun shine or the wind blow on demand (National Renewable Energy Laboratory [NREL], 2018).

This flips the traditional grid model on its head. Instead of adjusting supply to meet demand, we now need to either:

1. Store renewable energy when it's abundant for use when it's not, or

2. Adjust our electricity demand to match when renewable supply is available

It's like switching from a water system with a reliable tap to one where water randomly pours from the sky, and you need to figure out how to collect it, store it, and make it available whenever someone wants to take a shower.

This challenge is compounded by the fact that our best renewable resources are often located far from population centers. The sunniest deserts and windiest plains are typically nowhere near our cities. Imagine if all the grocery stores were suddenly relocated to remote mountaintops—we'd need a whole new system to get food to people.

The existing transmission system wasn't built to carry massive amounts of electricity from, say, solar farms in rural Arizona to power-hungry cities in California, or from wind farms in Iowa to Chicago. It's like trying to funnel a rushing river through a garden hose—the capacity simply isn't there (NREL, 2024a).

The Duck Curve: Not a Waterfowl Fashion Trend

One of the most illustrative challenges of integrating renewables into the grid is something engineers call the "duck curve"—which, disappointingly, has nothing to do with adorable waterfowl on a runway.

The duck curve represents the net electricity demand (total demand minus solar generation) over the course of a day in places with high solar penetration, like California. When graphed, the midday dip in net demand followed by a steep ramp-up in the evening creates a shape that vaguely resembles a duck (California Independent System Operator [CAISO], 2013).

Here's how it works:

1. **Morning**: Electricity demand rises as people wake up
2. **Midday**: Solar power floods the grid, reducing net demand for other power sources
3. **Evening**: Solar production drops as the sun sets, just as people come home and turn on appliances, causing a steep ramp-up in demand for other power sources

This creates two major challenges. First, conventional power plants may need to dramatically reduce output during midday solar production, then rapidly ramp up as the sun sets. Many traditional plants weren't designed for this roller coaster operation—it's like asking your grandparent to go from a nap to sprinting within five minutes.

Second, the duck curve gets more extreme as more solar is added to the grid. In places like California, midday solar production can sometimes exceed total demand, creating negative electricity prices (yes, that means producers literally pay consumers to use electricity). Meanwhile, the evening ramp-up becomes steeper and more challenging to manage (DOE, 2017).

This isn't just a theoretical issue. Grid operators in solar-heavy regions are already dealing with these challenges, and they'll only grow as we install more renewables. It's like throwing more ducks into a pond that wasn't designed for waterfowl—eventually, you're going to need a different pond. Or fewer ducks. Or duck-compatible pond renovations. I think I've stretched this metaphor to its breaking point.

Transmission Troubles: Powerline NIMBY

One of the biggest barriers to a renewable energy future isn't technology or even cost—it's getting permission to build the power lines needed to move clean electricity from where it's generated to where it's needed.

The United States needs to expand its transmission capacity by 60-90% by 2030 to support a clean energy transition, according to various analyses. That means building thousands of miles of new high-voltage transmission lines across the country (DOE, 2023a).

But here's the problem: nobody wants power lines in their backyard. Transmission projects face opposition from landowners, environmental groups concerned about wildlife impacts, local governments worried about property values, and competing business interests. It's like trying to build a new highway, except the highway is suspended in the air, carries invisible danger-juice, and makes an ominous buzzing sound.

The result is a bureaucratic nightmare. Major transmission projects can take 10+ years to complete, with endless permitting processes, environmental reviews, legal challenges, and regulatory approvals from multiple agencies at federal, state, and local levels. A single opponent can sometimes delay a project for years.

This transmission bottleneck is increasingly the binding constraint on renewable energy deployment. We have the technology to generate clean energy. We have the economic incentive to build it. We even have willing investors ready to finance it. What we don't have is a streamlined way to build the transmission lines needed to deliver that energy (NREL, 2024b).

It's a classic collective action problem: everyone benefits from a modernized grid with more clean energy, but the costs and inconveniences of new transmission infrastructure fall on specific communities and individuals. Finding fair ways to distribute these burdens is one of the central challenges of the energy transition.

Storage: The Holy Grail

If transmission is the backbone of a renewable grid, storage is its holy grail—the technology that could potentially solve the intermittency problem once and for all. After all, if we could cheaply store massive amounts of electricity, it wouldn't matter when the sun shines or the wind blows. We could just save the energy for when we need it, like squirrels hoarding nuts for winter, except with fewer bushy tails and more lithium (DOE, 2022).

Battery technology has improved dramatically in recent years, primarily driven by the electric vehicle revolution. Lithium-ion battery costs have fallen by nearly 90% since 2010, making grid-scale battery installations increasingly economical (NREL, 2021).

But we're still nowhere near having enough storage capacity to support a predominantly renewable grid. For context, the entire world currently has about 30 gigawatts of battery storage installed—enough to power the United States for...about 15 minutes. That's right: if we had to rely solely on current battery capacity during a cloudy, windless day, we'd be plunged into darkness faster than you can say "should have bought those emergency candles."

Beyond batteries, other storage technologies are emerging:

1. **Pumped hydro storage**: Using excess electricity to pump water uphill, then releasing it through turbines when energy is needed. It's like creating an artificial waterfall that only flows when you want electricity. This is currently the most common form of grid-scale energy storage, but it's limited by geography and environmental concerns (Environmental and Energy Study Institute [EESI], 2019).

2. **Compressed air energy storage**: Forcing air into underground caverns when electricity is abundant, then releasing it through turbines when needed. Think of it as bottling electricity in the form of pressurized air—like saving your breath to blow out birthday candles later, except significantly more useful.

3. **Thermal storage**: Storing energy as heat in materials like molten salt, which can later generate steam to drive turbines. It's

basically a giant thermos keeping your energy coffee hot until you're ready to drink it (DOE, 2022).

4. **Hydrogen**: Using excess renewable electricity to split water into hydrogen and oxygen, then recombining them in fuel cells when energy is needed. The hydrogen economy promises long-duration storage potential, but significant efficiency challenges remain. It's like converting your dollars to a foreign currency with a really bad exchange rate, then converting back when you need to spend them (International Hydropower Association, 2024).

The storage challenge isn't just about technology, but also about scale and economics. We need storage solutions that can economically shift energy across hours, days, and even seasons to fully rely on renewable energy. It's the difference between packing a snack for an afternoon outing versus storing enough food to last through winter.

Smart Grids: Teaching an Old Grid New Tricks

If storage is about shifting energy through time, the smart grid is about using information technology to make the entire system more efficient, flexible, and resilient—essentially teaching our dumb grid some new tricks.

The traditional power grid was designed for one-way communication: electricity flows from power plants to consumers, and utilities have limited visibility into what's happening at the distribution level. It's like trying to manage traffic in a city where you can only see the highways but none of the local streets.

A smart grid adds sensors, communications networks, and automated controls throughout the system, creating two-way flows of both electricity and information. This enables:

1. **Real-time monitoring**: Utilities can see exactly what's happening across the grid, identifying problems before they cascade into widespread outages. It's like giving the grid a full-body health monitoring system instead of waiting for something to obviously break.

2. **Automated response**: Software can automatically reroute power around damaged lines or balance supply and demand in milliseconds. Imagine if your home's plumbing could instantly redirect water around a burst pipe—that's what smart grids can do with electricity.

3. **Demand response**: Smart appliances and thermostats can automatically adjust electricity usage based on grid conditions or electricity prices. Your washing machine might decide to run at 2 AM when wind power is abundant and prices are low, or your electric vehicle might charge when the sun is shining rather than immediately when you plug it in (NREL, 2018).

4. **Distributed energy integration**: The smart grid can better manage two-way power flows when consumers also become producers through technologies like rooftop solar and home batteries. It turns the traditional grid model into something more like an energy internet, where everyone can both consume and contribute.

Smart grid technologies could substantially reduce the amount of new physical infrastructure needed for the clean energy transition. By using existing assets more efficiently and intelligently, we might be able to integrate more renewables with fewer new transmission lines—like finding a way to fit more cars on existing highways rather than building new ones.

However, smart grid development faces its own challenges, including cybersecurity concerns, privacy issues, regulatory barriers, and the sheer complexity of upgrading a century-old system while keeping the lights on. It's like trying to replace your car's engine while driving down the highway—tricky, to say the least.

Microgrids: Small Is Beautiful

While much of the grid modernization discussion focuses on continent-spanning transmission lines and massive wind farms, there's a parallel trend toward smaller, more localized systems called microgrids.

A microgrid is essentially a miniature version of the larger grid that can operate either connected to the main grid or independently (in "island mode"). Typically serving a specific geographic area like a university campus, hospital, military base, or remote community, microgrids combine local generation (often renewables plus backup generators), storage, and smart controls.

Microgrids offer several advantages:

1. **Resilience**: When the main grid fails during storms or other disasters, microgrids can keep critical facilities powered. It's like having your own lifeboat when the main ship is sinking.

2. **Efficiency**: By generating electricity close to where it's used, microgrids reduce transmission losses. It's the energy equivalent of growing your own vegetables instead of shipping them across the country.

3. **Renewable integration**: Microgrids can maximize local renewable resources and provide a testing ground for new technologies and control strategies. They're like renewable energy laboratories where we can experiment without risking the entire grid (DOE, 2024).

4. **Community control**: Microgrids can give communities more say in their energy decisions rather than depending entirely on distant utilities. Think of it as the locavore movement, but for electricity.

Microgrids are particularly valuable for remote areas where traditional grid connections are difficult or expensive. In Alaska, Hawaii, and island communities worldwide, microgrids are already proving that high renewable penetration is possible with the right mix of technologies.

However, microgrids aren't a complete solution to our grid challenges. They still rely on the larger grid for backup power in most cases, and they can't replace the economies of scale that come with larger systems. It's like growing some of your own food in a backyard garden—wonderful and meaningful, but probably not enough to completely replace the grocery store.

The future likely involves a hybrid approach—a more robust, interconnected macrogrid for efficient long-distance energy sharing, combined with more resilient, semi-autonomous microgrids at the local level. It's the electricity version of "think globally, act locally"—except with more transformers and fewer bumper stickers.

The Political Grid: Red, Blue, and Green All Over

If you thought the technical challenges of grid modernization were daunting, wait until you see the political obstacles. The grid isn't just a physical infrastructure—it's also embedded in a complex web of regulations, market structures, and political interests that can be even harder to change than the hardware itself.

The United States has a particularly fragmented system of grid governance. Different parts of the grid are regulated by the Federal Energy Regulatory Commission (FERC), state public utility commissions, regional transmission organizations (RTOs) and independent system operators (ISOs), plus municipal utilities and rural electric cooperatives that often have their own governance structures.

This patchwork creates a coordination nightmare for projects that cross jurisdictional boundaries—like, say, most of the transmission lines needed for a renewable energy future. It's like trying to plan a road trip through 50 countries, each with different driving rules, currency, and language (DOE, 2023b).

The politics get even messier when you consider that the clean energy transition creates both winners and losers. Communities built around fossil fuel production may resist changes that threaten their economic base, while utilities with large investments in existing infrastructure may be wary of disruptive technologies that could strand those assets. It's the energy equivalent of asking turkey farmers to enthusiastically promote Tofurky.

Different regions also have different clean energy priorities based on their available resources. The Southwest pushes for solar development, the Great Plains for wind, the Pacific Northwest for hydropower. Finding policies that work across these varied contexts is challenging, to say the least.

Despite these obstacles, grid politics is slowly evolving. States from California to New York have implemented ambitious clean energy standards that are driving grid investments. The federal government has allocated billions for grid modernization. And even traditionally conservative states like Texas and Iowa have become renewable energy powerhouses due to their abundant resources and economic benefits.

The clean grid transition ultimately needs to transcend partisan divisions —after all, everyone benefits from a more reliable, resilient, and clean electricity system. But getting there requires navigating complex economic transitions, balancing various interests, and making difficult tradeoffs. It's less about flipping a switch and more about rewiring our entire energy political economy while keeping the lights on.

Electrify Everything: The Grid's Expanding Job Description

As if modernizing the grid for renewable energy wasn't challenging enough, we're simultaneously planning to dramatically increase the amount of stuff that uses electricity. The clean energy transition doesn't just involve changing how we generate electricity—it also means switching other energy uses from fossil fuels to electricity. It's like asking someone to completely rebuild their car's engine while also expecting them to carry more passengers and drive faster.

This concept, often called "electrify everything," means replacing fossil fuel technologies with electric alternatives across the economy:

1. **Transportation**: Switching from gasoline and diesel vehicles to electric cars, trucks, and buses (DOE, 2022)
2. **Buildings**: Replacing gas furnaces, water heaters, and stoves with electric heat pumps and induction cooktops
3. **Industry**: Converting industrial processes from fossil fuels to electricity where possible (DOE, 2023c)

The logic is compelling: as the grid gets cleaner, electrification delivers increasing climate benefits. An electric car charged with coal power has a similar carbon footprint to a conventional vehicle, but one charged with wind or solar is dramatically cleaner. It's like how eating that kale salad only counts as healthy if you don't smother it in ranch dressing and croutons.

But here's the catch: electrification could potentially double or even triple electricity demand by mid-century. This means we're not just rebuilding the grid for today's needs—we're building for a future where electricity does much more work. It's like renovating your house while simultaneously planning to have triplets, start a home business, and take in your in-laws. The capacity requirements are, shall we say, substantial.

This electrification trend creates both challenges and opportunities for grid planners. On one hand, it means building much more generation, transmission, and distribution capacity than would otherwise be needed. On the other hand, many electric technologies—like electric vehicles and heat pumps—can be operated flexibly, potentially helping to balance variable renewable energy rather than exacerbating integration challenges.

For example, with smart charging, electric vehicles can recharge when renewable energy is abundant and electricity is cheap. Vehicle-to-grid technology could even allow EVs to feed electricity back to the grid during peak demand periods, essentially turning millions of car batteries into a massive distributed storage resource. It's like having a personal power plant in your garage, except it also takes you to Costco.

Similarly, grid-interactive water heaters and space heating systems can pre-heat when renewable energy is plentiful and then coast through periods of low renewable generation. The thermal mass of buildings themselves can become a form of energy storage, with smart thermostats adjusting temperatures slightly to shift electricity demand without compromising comfort.

The key insight is that electrification and grid modernization need to be planned together, not as separate initiatives. Electrification without clean electricity just shifts emissions from tailpipes to smokestacks. And renewable electricity without flexible demand makes grid integration unnecessarily difficult. It's a complex choreography that requires coordination across sectors that have historically operated independently.

Equity and Justice: Who Gets the Clean Grid First?

As we rebuild the grid for a clean energy future, we face important questions about who benefits from these investments and who bears the costs. The old fossil-fuel grid wasn't exactly a model of equity— pollution from power plants disproportionately affected low-income communities and communities of color, while rural and tribal areas often received less reliable service than wealthy urban centers.

The clean grid transition offers an opportunity to address these historical inequities, but only if equity and justice considerations are deliberately integrated into planning and investment decisions. Otherwise, we risk creating a two-tier system where affluent communities enjoy the benefits of clean, modern grid infrastructure while disadvantaged communities are left behind (Center for American Progress, 2022).

Several equity challenges deserve particular attention:

1. **Energy burden**: Lower-income households already spend a higher percentage of their income on energy. If the transition increases electricity rates without offsetting benefits, it could exacerbate this burden. It's like how a $5 latte is a trivial expense for some but a significant splurge for others (DOE, 2024).

2. **Grid reliability disparities**: Historically marginalized communities often experience more frequent outages and slower restoration times. As extreme weather increases with climate change, ensuring these communities benefit from grid resilience investments becomes even more important.

3. **Clean energy access**: Rooftop solar and other distributed energy resources have predominantly been adopted by higher-income households with suitable property and good credit. Expanding access to these technologies requires targeted programs and alternative ownership models.

4. **Just transition for workers**: As fossil fuel power plants close, the workers and communities that depended on them need support to transition to new opportunities. The clean energy

economy promises more jobs overall, but not necessarily in the same locations or for the same people.

Fortunately, policymakers and utilities are increasingly incorporating equity into their grid modernization efforts. Community solar programs make clean energy accessible without requiring home ownership. Energy efficiency initiatives target low-income households with the highest energy burdens. Workforce development programs prepare displaced fossil fuel workers for clean energy careers.

Some jurisdictions are going further by explicitly defining "energy justice" metrics and requiring that a percentage of clean energy investments benefit disadvantaged communities. New York's Climate Leadership and Community Protection Act, for example, requires that at least 35% of clean energy benefits flow to disadvantaged communities (Environmental Law & Policy Center, 2022).

These approaches recognize that the grid isn't just technical infrastructure—it's a social system that reflects and can either reinforce or help correct broader societal inequities. Building a cleaner grid isn't enough; we need to build a fairer one too.

What You Can Do: From Grid Complaints to Grid Solutions

All this grid talk might seem distant from your daily life—something for engineers, executives, and policymakers to worry about while you focus on more immediate concerns like whether you remembered to charge your phone. But the clean grid transition needs more than technical expertise and policy reform—it needs public understanding and support. After all, we're talking about rebuilding one of society's most fundamental systems during a climate emergency. That's not something a small group of experts can or should do alone.

Here are some ways you can contribute to a cleaner, more modern grid:

1. **Become an informed advocate**: Learn about your local electricity system and engage in public processes around grid investments. Showing up at utility commission hearings might not be your idea of a fun night out, but these forums often determine whether billions get invested in fossil fuel infrastructure or clean alternatives.

2. **Support transmission expansion**: When new transmission projects are proposed in your region, consider the broader climate benefits rather than just local impacts. NIMBYism is one of the biggest barriers to the clean grid transition. Yes, power lines aren't exactly scenic, but neither is climate change.

3. **Be a flexible electricity consumer**: If you have the option, enroll in time-of-use electricity rates or demand response programs that incentivize using power when clean energy is abundant. Run your dishwasher and charge your EV when the sun is shining or the wind is blowing.

4. **Invest in distributed energy where appropriate**: If you own your home and have suitable roof space, consider solar panels, home batteries, or other clean energy technologies. These not only reduce your carbon footprint but also build resilience against grid outages.

5. **Electrify thoughtfully**: When replacing appliances or vehicles, choose electric options where possible, but pair them with smart

controls that can respond to grid conditions. An electric vehicle with managed charging is an asset to the clean grid; one that always charges immediately at 6 PM during peak demand is a potential problem.

6. **Support equitable grid policies**: Advocate for programs that ensure low-income communities and communities of color share in the benefits of grid modernization. A clean grid that exacerbates existing inequities isn't truly sustainable.

7. **Vote for climate-aware candidates**: Ultimately, many grid decisions are shaped by elected officials who appoint utility commissioners, determine infrastructure investments, and set energy policies. The ballot box is a powerful tool for accelerating the clean grid transition.

Remember, the grid exists to serve human needs—not the other way around. As we rebuild this system for a clean energy future, your voice matters in determining what gets built, who benefits, and how quickly we transition. The grid might not be ready for the clean energy revolution, but with enough public engagement and political will, it can get there faster than the cynics expect.

Conclusion: Grid of Dreams

In the 1989 film "Field of Dreams," a mysterious voice tells Kevin Costner's character: "If you build it, he will come." The clean grid transition requires a similar leap of faith—except the voice is climate science, and it's telling us: "If you don't build it, we're all screwed."

The grid challenges we've explored in this chapter aren't just technical puzzles—they're challenges of imagination, coordination, and collective action. We need to build a grid for the future we want, not cling to the fossil-fueled past we know. It's like how you don't refuse to build more schools just because the current ones are full; you plan for the education system you need, not the one you have (NREL, 2022).

This transition involves a thousand complex decisions: where to build transmission lines, how to balance local resilience with system-wide efficiency, who pays for infrastructure that benefits everyone, how to ensure technology deployment doesn't worsen inequities. These aren't easy questions, but they're answerable if we approach them with clarity about our goals and commitment to finding solutions that work for all communities.

The good news is that the clean grid transition is already happening, driven by the plummeting costs of renewable energy, the increasing risks of climate change, and the tireless work of engineers, policymakers, activists, and ordinary citizens who understand what's at stake. Every solar panel installed, every transmission line approved, every battery deployed, and every smart meter activated brings us closer to a grid that can deliver clean, affordable, reliable electricity to everyone (DOE, 2023a).

Will it be easy? No. Will we encounter setbacks and failures along the way? Certainly. But the alternative—continuing to rely on a grid designed for fossil fuels in a world that desperately needs to eliminate carbon emissions—is far worse. The grid isn't ready for the clean energy revolution, and neither are we. But ready or not, the future is coming. We can either build the grid we need or suffer the consequences of climate inaction.

In the next chapter, we'll explore how sustainability starts at home—the actions individuals and communities can take to reduce their climate impact while broader systems change unfolds. Because while we need a clean grid to solve climate change, we can't afford to wait for it to be completed before starting our personal climate journeys.

Chapter 11: Your Backyard Matters: Sustainability Starts at Home

Getting Personal with Climate Change

Congratulations! You've made it through ten chapters of increasingly alarming climate science, complicated carbon cycles, and grid infrastructure challenges. By now you might be thinking, "OK, I get it—we're in trouble. But what am I supposed to do about it? I'm just one person with a Netflix addiction and an unreliable compost bin, not the CEO of ExxonMobil."

It's a fair question. When faced with a planetary problem as vast as climate change, individual actions can seem as effective as trying to put out a forest fire with a water pistol. Why bother changing your lifestyle when the real culprits are giant corporations, dysfunctional governments, and that neighbor who leaves their Christmas lights up until April?

Here's the uncomfortable truth: while system-level change is absolutely necessary (and we'll talk about how to push for that too), your personal choices do matter. Not just because they directly reduce carbon emissions—though they do—but because they send market signals, shift cultural norms, and build momentum for broader changes. It's like voting in an election where every day is Election Day, and your ballot is your wallet, your habits, and your conversations (Weber and Matthews 2008).

This chapter explores how sustainability starts at home—the personal and community-level actions that make a meaningful difference in addressing climate change. We'll cover everything from the food on your plate to the energy powering your doomscrolling habit, all while acknowledging that perfection isn't the goal and that systemic barriers make some "green" choices inaccessible to many. So grab your reusable water bottle (or whatever container you happen to have handy—no judgment here) and let's get personal about climate action.

Carbon Footprints: Size Does Matter

Before diving into specific actions, it's helpful to understand where your personal emissions come from. Your "carbon footprint" represents the total greenhouse gases emitted as a result of your lifestyle and choices. It's like your climate impact resumé, except nobody's impressed by a bigger number (Center for Sustainable Systems 2022).

For the average American, the major categories look something like this:

1. **Transportation**: ~28% (primarily from driving and flying)
2. **Home energy use**: ~17% (electricity and heating/cooling)
3. **Food**: ~14% (especially meat and dairy)
4. **Stuff you buy**: ~26% (all those Amazon packages add up)
5. **Services you use**: ~15% (everything from healthcare to entertainment)

These proportions vary significantly based on where and how you live. Someone in a walkable city with good public transit naturally has a lower transportation footprint than a rural resident who drives 50 miles to work. A person in Maine needs more heating than someone in Florida, who in turn uses more air conditioning. Climate action isn't one-size-fits-all, and what works in Brooklyn might not work in Boise.

That said, some choices have outsized impact across almost all lifestyles. These "carbon bigfoots" include:

- **Flying**: A single round-trip flight from New York to London generates roughly 1.8 metric tons of CO_2—nearly 10% of the average American's annual carbon footprint in one weekend trip. It's the climate equivalent of eating 800 cheeseburgers in a weekend, except with less intestinal distress (Perch Energy 2024).

- **Car dependency**: Driving a typical gasoline car 12,000 miles per year produces about 4.6 metric tons of CO_2. Electric vehicles cut this dramatically, but the best option remains driving less altogether (EPA 2024).

- **Meat consumption**: Beef production generates roughly 60 kg of greenhouse gases per kg of meat—about 10 times more than chicken and 30 times more than lentils. Your burger habit might be more climate-significant than your light bulbs (Poore and Nemecek 2018).

- **Home energy waste**: The average American home leaks enough air that it's equivalent to having a window open year-round. Improving insulation and sealing leaks can reduce heating and cooling emissions by 15-30% (U.S. Department of Energy 2023).

The good news is that focusing on these high-impact areas gives you the most climate bang for your buck. It's like how targeting belly fat gives you better health results than obsessing over your pinky toe weight. Prioritization matters.

Getting Around: Transportation Without the Carbon Baggage

Unless you're currently living in a cave (in which case, impressive internet connection!), transportation is likely a significant chunk of your carbon footprint. Americans drive over 3 trillion miles annually—enough to travel to Pluto and back about 16,000 times, though most of us are just commuting to jobs we don't particularly like.

The climate impact of your transportation choices follows a clear hierarchy:

1. **Walking and biking**: Near-zero emissions (unless you count the extra cheeseburger you eat to fuel your bike ride)
2. **Public transit**: Way lower per-person emissions than driving alone
3. **Electric vehicles**: Significantly better than gas cars, especially as the grid gets cleaner
4. **Efficient gas vehicles**: Better than gas-guzzlers, but still problematic
5. **Flying**: Particularly bad for the climate, especially long distances
6. **Cruise ships**: The floating carbon catastrophes of the transportation world

This hierarchy suggests an obvious approach: walk, bike, or take transit when possible; choose electric vehicles when not; and rethink your long-distance travel habits. But let's be real—transportation choices are constrained by infrastructure, economics, and life circumstances. Not everyone can bike to work or afford an electric car.

So what practical steps can you take, regardless of your situation?

Drive less, if you can: Combining errands, carpooling, working from home (even part-time), and choosing close-to-home options all reduce driving emissions. Even cutting just one car trip per week makes a difference over time—it's the climate equivalent of a modest diet rather than extreme fasting (EPA 2024).

Make your next vehicle electric: Electric vehicles (EVs) produce about half the lifetime emissions of comparable gas cars today, and this advantage grows as the grid gets cleaner. The upfront cost remains higher, but lower operating costs and increasing incentives are closing the gap. Plus, never having to visit a gas station again is its own reward (U.S. Energy Information Administration 2023).

Rethink air travel: For many of us, flying is the largest single source of emissions. Consider whether some trips could be replaced with virtual connections or alternate transportation. When you do fly, direct flights at high capacity create fewer emissions than connecting flights on half-empty planes. And despite what your meditation app says, you don't actually *need* to visit Bali to find inner peace.

Support better transportation infrastructure: Even if your personal options are limited now, supporting investments in public transit, bike lanes, pedestrian safety, and EV charging helps create systems where low-carbon transportation is viable for more people. Climate-friendly infrastructure doesn't appear without advocacy and political support.

The transportation transition involves both individual choices and system changes. Each electric vehicle purchased helps build market demand for more models at better prices. Each person choosing to bike or take transit demonstrates community support for these options. Your choices matter not just for their direct impact, but for the signals they send and the momentum they build.

Home Sweet (Sustainable) Home

Your home is your castle, your sanctuary, your place to walk around in underwear without judgment. It's also, for most Americans, a significant source of greenhouse gas emissions through energy use for heating, cooling, lighting, and powering the growing collection of devices that beep at inconvenient hours.

The good news is that making your home more sustainable often saves money while increasing comfort. It's one of the rare win-win-win situations in climate action, benefiting your wallet, your well-being, and the planet simultaneously. Here's how to approach home sustainability, roughly in order of impact:

Electrify everything (eventually): As the grid gets cleaner, replacing fossil fuel appliances with electric alternatives dramatically reduces your home's carbon footprint. When your gas furnace, water heater, or stove reaches the end of its life, consider electric heat pumps and induction cooking. This transition can happen gradually as equipment needs replacement—no need to throw out functioning appliances (International Energy Agency 2022).

Weatherize and insulate: Before investing in fancy new heating and cooling systems, make sure your home isn't leaking energy like a sieve. Proper insulation, air sealing, and weather stripping are the unglamorous workhorses of home sustainability. It's like putting on a sweater before cranking up the heat—basic, effective, and fiscally responsible.

Choose efficient appliances: When it's time to replace appliances, look for energy-efficient models with ENERGY STAR certification. The efficiency of refrigerators, washing machines, and other major appliances has improved dramatically over the decades. That harvest gold fridge from 1975 might have retro charm, but its energy usage is also retro—and not in a good way.

Consider rooftop solar: If you own your home and have suitable roof space, solar panels can reduce both your carbon footprint and your electricity bills. Installation costs have fallen dramatically while efficiency has improved. Various financing options make solar accessible with little or no upfront cost, though the specifics vary by location (Rocky Mountain Institute 2023).

Optimize with smart technology: Programmable thermostats, smart power strips, and home energy monitors help eliminate waste without sacrificing comfort. They're like having a frugal roommate who actually remembers to turn things off, except without the passive-aggressive notes about your shower length.

Choose renewable electricity: Many utilities offer options to source your electricity from renewable energy, typically for a small premium. Community solar programs provide another pathway to clean energy, often without requiring suitable roof space or home ownership (NREL 2024).

The emissions impact of these actions varies widely based on your current home, climate, and local energy sources. In general, focus first on the biggest energy users—typically heating, cooling, and water heating—before obsessing over whether your phone charger is drawing phantom power. It's like dieting by cutting out daily donuts rather than agonizing over the occasional breath mint.

For renters and those with limited resources, options may be more constrained but still exist. Energy-efficient lighting, smart power strips, window insulation kits, and programmable thermostats offer affordable improvements that often pay for themselves quickly. And advocating for better building codes and landlord efficiency requirements helps ensure that housing at all price points becomes more sustainable over time.

You Are What You Eat (Including Climatically)

Food might not seem like an obvious climate issue compared to SUVs and coal plants, but agriculture generates about a quarter of global greenhouse gas emissions. Your dietary choices have a significant climate impact—though exactly how significant depends on what and how you eat.

The carbon footprint hierarchy for food is clear but nuanced:

1. **Red meat (especially beef)**: Far higher emissions than almost any other food due to methane from cow digestion and deforestation for grazing and feed crops
2. **Dairy**: Lower than beef but still high, for similar reasons
3. **Other animal products**: Pork, chicken, eggs, and fish have substantially lower footprints than beef but higher than plant foods
4. **Plant foods**: Generally much lower carbon footprints, though transportation and production methods matter

This hierarchy suggests an obvious approach—eat less beef and dairy—but fully embracing plant-based eating isn't feasible or desirable for everyone. The good news is that you don't need to go fully vegan to make a significant difference. Even reducing meat consumption or switching from beef to chicken can substantially cut your dietary carbon footprint (Ritchie 2023).

Beyond what you eat, how your food is produced matters too:

Food waste: About 30-40% of food in the United States is wasted, generating emissions throughout the supply chain for food that never gets eaten. Better meal planning, proper food storage, creative use of leftovers, and composting inedible scraps all help reduce this massive inefficiency (United Nations 2023).

Local and seasonal eating: While "food miles" get a lot of attention, transportation is actually a relatively small part of most foods' carbon footprint compared to production methods. Local food can have benefits for freshness, community economics, and connection to food sources, but don't assume local automatically means lower carbon. A tomato grown in a heated greenhouse in winter might have a higher footprint than one shipped from a region where it grows naturally.

Regenerative agriculture: How food is grown matters tremendously for its climate impact. Regenerative practices like cover cropping, reduced tillage, and holistic grazing can potentially sequester carbon in soil while producing nutritious food. Supporting farmers using these methods through farmers markets, CSAs (Community Supported Agriculture), and conscientious purchasing helps expand their adoption.

The key takeaway? You don't need dietary perfection—just progress. If everyone in the United States ate chicken instead of beef just one day per week, it would have the climate benefit of taking millions of cars off the road. Small, consistent changes across many people add up to significant impact.

And remember, food choices are deeply personal, cultural, and emotional. Climate considerations are important, but they're not the only valid factors in deciding what to eat. Finding a sustainable approach that works with your circumstances, values, health needs, and cultural traditions is what matters in the long run.

Stuff: Because You Probably Have Too Much Already

Let's be honest: Most of us living in wealthy countries have too much stuff. Our closets overflow with clothes we rarely wear. Our garages are so packed with possessions that our cars live on the driveway. We rent storage units to hold the excess that won't fit in our homes. And all this stuff carries a significant carbon footprint from resource extraction, manufacturing, transportation, and eventual disposal (Taiebat and Xu 2019).

The climate impact of our material consumption habits is enormous but often overlooked. While energy and transportation get most of the attention, the things we buy account for a substantial portion of our carbon footprints. Every new item requires energy and resources to produce, typically generating greenhouse gas emissions in the process.

Fortunately, reducing your "stuff footprint" doesn't mean living like an ascetic monk (unless that's your thing—no judgment). Instead, try these approaches:

Buy less, choose well: Before purchasing something new, ask whether you really need it, whether it will last, and whether you'll still want it in a year. Investing in fewer, higher-quality items often saves money in the long run while reducing environmental impact. Marie Kondo had a point about only keeping things that "spark joy," though to be fair, my pizza cutter doesn't exactly fill me with delight—it just does its job reliably.

Embrace the circular economy: Instead of the linear take-make-waste model, look for opportunities to participate in circular systems where resources are kept in use. This includes buying used items, repairing broken things, repurposing what you have, and ensuring recyclability at end of life. It's like giving your stuff multiple lives instead of a one-way ticket to landfill purgatory (Greenly 2024).

Share rather than own: Many items that we use infrequently don't need to be personally owned. Tool libraries, toy lending programs, formal and informal neighborhood sharing systems, and rental services can provide access without the environmental impact (or cost) of individual ownership. Your power drill will spend 99% of its life sitting unused anyway—might as well let the neighbors benefit during its extensive downtime.

Consider embedded emissions: Some products have surprisingly large carbon footprints due to energy-intensive manufacturing processes or materials. Electronics, concrete, aluminum, and new clothing often fall into this category. Being aware of these "carbon hotspots" helps direct your attention to high-impact purchasing decisions.

This doesn't mean you should never buy anything new or feel guilty about every purchase. Rather, it's about approaching consumption more mindfully and finding a sustainable balance that works for your life. And importantly, these approaches often save money while reducing stress and clutter—climate benefits aside, having less stuff can simply make life more pleasant.

A special note about fast fashion: The clothing industry has a particularly problematic environmental footprint, with fast fashion driving ever-more-rapid cycles of purchase and disposal. The average

American now buys about five times as many clothing items as in 1980, while wearing each item far fewer times. Breaking this cycle through more durable clothing choices, secondhand shopping, clothing swaps, or rental services can significantly reduce your carbon footprint while potentially improving your style game. Vintage is cooler anyway, right?

Money Talks: Financial Choices for Climate Impact

Your money doesn't just sit inert in bank accounts and investment portfolios—it's actively at work in the world, potentially funding either climate solutions or climate problems. Financial choices might seem abstract compared to tangible actions like biking to work, but they can have equal or greater climate impact (McKinsey 2022).

Here's how to align your money with your climate values:

Banking with climate in mind: Major banks fund fossil fuel projects with your deposits. Since 2016, the world's 60 largest banks have provided over $3.8 trillion in financing to fossil fuel companies. Consider moving your accounts to banks, credit unions, or online institutions with stronger climate commitments. It's one of the easiest climate actions with potentially significant impact.

Climate-conscious investing: If you have investments, you're potentially a part-owner in companies affecting climate change—for better or worse. Options for climate-aligned investing include:

- Fossil-free funds that exclude the most problematic industries
- ESG (Environmental, Social, Governance) funds that consider sustainability factors
- Thematic investments specifically focused on climate solutions

- Shareholder activism to push existing companies toward better practices

You don't need to be a millionaire to make a difference—even small retirement accounts or college savings plans can be directed toward climate-friendly options.

Insurance considerations: Insurance companies both protect against climate risks and invest enormous sums in various industries. Some insurers have begun restricting coverage for fossil fuel projects while increasing investments in clean energy. Research your providers' climate positions and consider companies with stronger sustainability commitments.

Charitable giving: If you're able to donate to charitable causes, consider supporting organizations working on climate solutions—from policy advocacy to direct emissions reduction to climate justice initiatives. Research suggests that high-impact climate charities can prevent a ton of CO_2 emissions for as little as $1-10 in donations, making charitable giving potentially one of the most cost-effective climate actions (Food Tank 2023).

These financial choices matter not just for their direct impact, but also for the market signals they send. When banks, investors, and insurers see customers making decisions based on climate considerations, they respond with better policies and products. Your money has a voice— make sure it's saying what you want about the future you're hoping to create.

And a final note: While we often focus on the "don'ts" of climate finance (don't fund fossil fuels), the "dos" are equally important.

Financing climate solutions like renewable energy, sustainable agriculture, and clean transportation helps accelerate the transition we urgently need. Your money can either prop up the carbon-intensive past or help build the sustainable future—the choice is yours.

Beyond Individual Actions: Community Climate Solutions

While personal choices matter, climate action gets more powerful when we act collectively. Community-scale solutions amplify impact, address systemic barriers, and create space for broader participation. Plus, they're often more fun—because saving the world is better with friends (Climate Resolve 2023).

Here are some ways to engage beyond your household:

Community solar: Can't install panels on your own roof? Community solar allows multiple households to share the benefits of a larger solar installation, often with no upfront cost and immediate savings on electricity bills. These programs make clean energy accessible to renters, condo dwellers, those with unsuitable roofs, and people with limited financial resources (NREL 2024).

Food systems transformation: Farmers markets, community gardens, food co-ops, CSA programs, and urban agriculture initiatives all help build more sustainable local food systems while strengthening community connections. Growing even a small portion of your food locally reduces emissions while providing numerous co-benefits for health, security, and community resilience.

Sharing economy initiatives: Tool libraries, repair cafés, clothing swaps, timebanks, and community workshops reduce consumption while building skills and relationships. These approaches recognize that we don't all need to individually own rarely-used items when we can share resources within our communities.

Transportation advocacy: Supporting public transit, bike infrastructure, walkable development, and EV charging networks helps create systems where sustainable transportation choices are viable for more people. Even if you personally can't abandon your car today, advocating for better options helps build the infrastructure needed for broad-based change.

Community energy efficiency: Programs like weatherization assistance, bulk purchasing of efficient appliances, and neighborhood energy challenges help overcome barriers to home energy improvements. These approaches are particularly important for ensuring that lower-income households can participate in and benefit from the clean energy transition.

Climate-friendly land use: Protecting and restoring natural carbon sinks through community forests, wetland preservation, urban tree planting, and regenerative land management helps sequester carbon while providing wildlife habitat, flood mitigation, urban cooling, and recreational opportunities (United Nations 2022).

These community approaches address limitations of individual action by creating economies of scale, overcoming structural barriers, and developing models that work for diverse circumstances. They recognize that not everyone can afford solar panels or electric vehicles as individuals, but collectively we can create systems that make sustainable choices accessible to all.

Importantly, community climate action builds social capital and resilience alongside environmental benefits. The relationships, skills, and networks developed through these initiatives strengthen our capacity to navigate both climate impacts and the clean energy transition. It turns out that the "sustainability" we need isn't just about carbon accounting —it's also about building communities that can adapt and thrive through changing conditions.

Climate Citizenship: Because Policies Matter Too

Even the most committed individual and community climate actions can't fully address a global challenge without supportive policies. While changing lightbulbs and eating less meat are important, changing laws and policies often has greater leverage. That's where climate citizenship comes in—using your voice and vote to support climate action at all levels of government (Feldman and Hart 2021).

Here's how to be an effective climate citizen:

Vote climate: In every election from local to national, research candidates' climate positions and make this a priority in your voting decisions. Climate policy has become increasingly partisan in the United States, but effective climate voters look beyond party labels to specific commitments and records. And don't underestimate local elections— mayors, city councils, and county commissioners make crucial decisions about transportation, buildings, land use, and energy that significantly impact emissions.

Engage with representatives: Elected officials respond to constituent priorities. Calls, emails, letters, and in-person visits signal that climate matters to voters. Research suggests that relatively few constituent contacts can significantly influence representatives' positions, especially on issues they don't hear about often. Your voice has more impact than you might think, particularly at local and state levels where officials receive less communication overall.

Support climate ballot measures: Many jurisdictions allow direct democracy through ballot initiatives on specific policies. Climate-related measures might address renewable energy standards, transportation funding, conservation, or other issues. These direct votes can accelerate climate action even when elected officials lag behind public opinion.

Engage in public processes: Beyond elections, numerous opportunities exist to shape climate-relevant decisions. Public comments on proposed regulations, participation in utility resource planning, testimony at zoning hearings, and engagement with climate planning processes all help strengthen climate policies. These venues might seem dry (and they often are), but they're where consequential decisions get made (Citizens' Climate Lobby 2024).

Connect climate to other issues: Climate change intersects with almost every other social concern, from public health to economic justice to national security. Helping others see these connections builds broader support for climate action. Rather than treating climate as a standalone environmental issue, highlight how climate solutions can help address other priorities your community cares about.

Find your lane: Effective climate citizenship leverages your specific skills, knowledge, and relationships. Teachers can integrate climate into education; healthcare providers can highlight climate-health connections; business owners can implement sustainable practices; faith leaders can frame climate as a moral issue; artists can inspire through creative expression. Whatever your position and talents, there's a way to advance climate action that fits your authentic role in your community.

Climate citizenship recognizes that we're not just consumers making market choices—we're also citizens with voices in collective decisions. The most impactful climate actions often involve a combination of personal practices that align with your values, community engagement that builds alternatives and demonstrates what's possible, and policy advocacy that creates enabling conditions for broader transformation.

And importantly, engaging as a climate citizen helps counter the helplessness that can come from facing such an enormous challenge. Taking concrete actions alongside others, even when the immediate results aren't world-changing, builds agency and hope. It's like the difference between watching a depressing documentary about a problem versus actually joining a movement working on solutions—both might be sobering, but only one leaves you feeling like part of the answer.

Finding Your Climate Action Sweet Spot

With so many possible climate actions—from dietary changes to political engagement—how do you decide where to focus your limited time and energy? It can feel overwhelming, like being handed a 50-page menu when you're already hungry. Fortunately, you don't need to do everything to make a meaningful difference (Rewiring America 2024).

The most effective approach combines these considerations:

High-impact actions: Focus first on changes that significantly reduce emissions rather than minor tweaks. Transportation, home energy, and food choices typically offer bigger carbon savings than obsessing over plastic straws or standby power. It's like dieting by reducing daily donuts rather than agonizing over whether your multivitamin has an extra calorie.

Leverage points: Some actions have cascading effects beyond their direct emissions impact. Installing visible solar panels normalizes clean energy in your neighborhood. Supporting a local climate policy could reduce thousands of times more emissions than personal lifestyle changes. Speaking about climate from your authentic perspective might influence people who wouldn't listen to traditional environmentalists.

Personal circumstances: Your geographic location, housing situation, income, family responsibilities, and other factors shape which actions are feasible. Someone in a rural area without public transit faces different opportunities than someone in a walkable city. Working with rather than against your circumstances leads to more sustainable changes.

Co-benefits and joy: Climate actions that also improve your life in other ways are more likely to stick. Biking instead of driving improves health while saving money. Growing some of your own food connects you to natural cycles while providing fresh produce. A smaller home with less stuff can reduce both emissions and stress. Look for these win-win opportunities.

Skills and interests: Leveraging your existing talents and passions makes climate action more authentic and sustainable. If you love cooking, focus on low-carbon cuisine. If you're a natural organizer, build community climate initiatives. If you're technically inclined, explore home energy solutions. Your climate contribution shouldn't feel like an obligation separate from your identity—it works best as an expression of who you already are.

This framework helps identify your climate action "sweet spot"—the intersection of high impact, personal circumstances, co-benefits, and authentic expression. This sweet spot is different for everyone, and it shifts over time as your situation and broader systems change. The climate actions that make sense for you today might not be the same ones that make sense five years from now, and that's completely fine.

The key is getting started with actions that feel meaningful and sustainable for your situation, rather than striving for an impossible perfect standard. Climate action isn't about moral purity or personal virtue—it's about contributing to a massive global transition away from fossil fuels. Every step in that direction matters, even if it seems small in isolation.

And remember that climate actions tend to build on each other. The person who installs solar panels today might become an advocate for community renewable energy tomorrow. The family that tries plant-based meals might eventually support sustainable agriculture policies. Your climate journey doesn't need to start with the most ambitious actions—it just needs to start somewhere and keep evolving.

Conclusion: The Climate Adds Up

We've covered a lot of ground in this chapter, from transportation and home energy to food, consumption, finance, and beyond. It might seem overwhelming at first—so many possible actions across so many domains of life. But remember, you don't need to transform everything overnight. Climate action is a journey, not a destination, and every step reduces harm while building momentum toward broader change (Robert Wood Johnson Foundation 2024).

The amazing thing about climate action is how it adds up—across actions, across people, and across time:

Across actions: Each choice might seem small in isolation, but collectively they reshape your relationship with carbon. The bike commutes, plant-based meals, energy retrofits, and community engagements combine to substantially reduce your footprint while modeling a different way of living within planetary boundaries.

Across people: Your individual impact multiplies when others are inspired to make similar changes. Cultural norms shift through social influence, not abstract information. By living your values and sharing your journey (without preaching or judging), you help normalize climate-friendly choices within your social circles.

Across time: Climate actions build momentum and capabilities over months and years. The person who starts with a few meatless meals might eventually advocate for sustainable food policies. The household that installs LED bulbs might later go all-electric and solar-powered. Initial steps create familiarity and confidence for more ambitious changes.

This cumulative perspective helps counter the despair that can come from comparing your individual actions to the enormous scale of the climate challenge. No, you alone can't solve climate change—but you're not alone, and your actions aren't happening in isolation. You're part of a growing global movement of people transforming how we live, work, eat, move, and relate to the natural world.

And importantly, climate action isn't just about sacrifice and restriction —it often improves quality of life in tangible ways. More connected communities. Healthier ecosystems. Cleaner air and water. More equitable access to resources and opportunities. Less traffic and noise. More connection to natural rhythms and local places. The sustainable future we're creating isn't just lower-carbon—it's potentially more satisfying, healthy, and just than the high-carbon status quo.

So start where you are, with actions that make sense for your circumstances and values. Be gentle with yourself when you fall short of your aspirations—perfection isn't the goal, and guilt doesn't reduce emissions. Celebrate progress rather than obsessing over missteps. Connect with others on similar journeys to share ideas, encouragement, and the occasional climate-friendly meal.

Your backyard really does matter in addressing climate change—not just the literal space behind your home, but your household, your daily habits, your community engagement, and your voice as a citizen. It matters because it's where abstract climate concern becomes concrete climate action. It matters because it's where you translate values into practices that influence others. And it matters because countless backyards, taken together, can transform the landscape of possibilities for our shared climate future.

In the next chapter, we'll explore how the climate is changing us—the psychological, social, and cultural dimensions of the climate crisis and how we might grow through this challenging moment in human history. Because climate change isn't just about what's happening to the atmosphere—it's also about what's happening to us as we face this unprecedented challenge.

Chapter 12: The Climate Is Changing --- Are You?

Inside the Climate Mind

We've spent eleven chapters exploring the science, impacts, and solutions to climate change. We've dissected carbon cycles, scrutinized melting ice sheets, tracked rising temperatures, and analyzed renewable energy transitions. But there's one critical element we haven't fully examined yet: the squishy three-pound organ between your ears that's trying to make sense of it all.

The human mind isn't exactly optimized for handling planetary-scale, slow-moving, probabilistic threats like climate change. Our brains evolved to respond to immediate dangers---like that rustling in the bushes that might be a lion---not gradual shifts in atmospheric composition that unfold over decades. It's like trying to use a hammer to fix a computer; the tool just wasn't designed for the job (Gifford, 2011).

This cognitive mismatch helps explain why climate change is so psychologically challenging. It's not just the enormity of the threat that makes it difficult to process---it's that our mental hardware struggles with threats that don't trigger our evolved fight-or-flight responses. No wonder so many of us vacillate between apocalyptic doom-scrolling and blissful denial, with precious little productive engagement in between.

In this chapter, we'll explore the psychology of climate change---why it's hard to grasp, why we resist change, and how to overcome these barriers. We'll examine cognitive biases, risk perception, values, and communication strategies that shape our climate responses. And we'll look at approaches that help us engage more effectively with this defining challenge of our time, both individually and collectively (Clayton et al., 2023).

Because ultimately, addressing climate change isn't just about changing energy systems, transportation networks, or agricultural practices---it's also about changing minds, including our own. The climate is changing---the question is whether we can change fast enough to meet this unprecedented challenge.

Climate Change and the Brain: A Design Mismatch

Before diving into specific psychological barriers, it helps to understand why climate change is such a perfect storm for human cognitive limitations. It's not that we're stupid (well, not always); it's that climate change has characteristics that our brains are particularly ill-equipped to handle.

Here's why climate change is a psychological perfect storm:

It's gradual: Our attention systems evolved to detect sudden changes, not slow shifts. A 1°C temperature increase over decades doesn't trigger our threat detection systems the way a sudden loud noise does. It's like the proverbial frog in slowly heating water (though real frogs actually do jump out---they're smarter than the metaphor gives them credit for). Our perceptual systems simply aren't designed to notice gradual changes without scientific instruments and careful record-keeping (Gifford, 2011).

It's complex: Climate science involves systems thinking, probability, and multi-causal relationships---all areas where human cognition struggles. We prefer simple, linear explanations with clear cause-and-effect relationships. "It's hot today because of climate change" is both technically incorrect and cognitively satisfying in a way that "climate change increases the probability of heat extremes through multiple feedback mechanisms" never will be (IPCC, 2022).

It's abstract: Carbon dioxide is invisible. Global temperature anomalies are statistical constructs. Climate models are mathematical abstractions. None of these trigger the concrete, sensory-based thinking that comes most naturally to humans. We're much better at responding to tangible threats we can see, hear, touch, or smell than to concepts that require abstract reasoning.

It's distant: Until recently, many people perceived climate change as temporally distant (affecting future generations), geographically distant (affecting other countries), and socially distant (affecting other species or people unlike themselves). Our cognitive and emotional systems discount such psychologically distant threats, making them feel less urgent than immediate, personal concerns. It's why we'll rush to the doctor for chest pain but postpone preventative screenings.

It's uncertain: While the fact of human-caused warming is scientifically certain, the precise timing and magnitude of specific impacts retain uncertainty ranges. Human cognition craves certainty and struggles with probabilistic thinking. We tend to either demand impossible precision ("Tell me exactly what will happen in my town in 2050") or use uncertainty as an excuse for inaction ("Scientists aren't 100% certain, so why worry?") (IPCC, 2022).

It's politically charged: In some countries, particularly the United States, climate change became entangled with political identity, transforming scientific facts into tribal signifiers. Once this happens, cognition shifts from accuracy-oriented to identity-protective.

Information gets processed not for its truth value but for its consistency with group beliefs. It's like how sports fans can watch exactly the same play and see completely different fouls depending on which team they support (Hornsey, 2021).

These characteristics create a "perfect storm" for our cognitive limitations. It's not that humans can't understand climate change---clearly many do---but that it requires overcoming significant psychological hurdles that don't exist for more immediate, concrete threats. Understanding these inherent challenges helps explain why simply providing more information often fails to change minds or motivate action.

Cognitive Biases: Your Brain's Climate Blind Spots

Beyond the general mismatch between climate change and human cognition, specific cognitive biases---systematic errors in thinking---further complicate our climate responses. These mental shortcuts served our ancestors well on the savanna but lead us astray when facing complex global challenges. They're like evolutionary software bugs that nobody's bothered to patch.

Here are some particularly problematic biases for climate change:

Present bias: We heavily discount future consequences compared to present costs or benefits. A dollar today feels more valuable than two dollars in ten years. Similarly, immediate convenience outweighs future climate impacts in our decision-making. It's why we'll choose the gas-guzzling convenience of driving over alternatives, even when we understand the long-term climate implications (Korteling et al., 2023).

Status quo bias: We have a strong preference for things as they are, perceiving changes as losses even when they're objectively beneficial. Shifting from familiar fossil fuel systems to new clean energy approaches triggers status quo bias even when the change would improve outcomes. It's like sticking with a dated phone operating system because it's familiar, despite the new one being objectively better.

Single-action bias: After taking one climate-friendly action, we often feel we've "done our part" and become less motivated to take additional steps. Installing LED bulbs might reduce motivation for more impactful actions like reducing air travel or advocating for climate policies. It's the cognitive equivalent of eating one salad and considering your nutrition goals accomplished for the year (Zhao & Luo, 2021).

Confirmation bias: We seek information that confirms our existing beliefs while ignoring contradictory evidence. Climate skeptics find and remember the rare studies questioning climate consensus while dismissing the overwhelming evidence supporting it. Climate advocates might similarly overestimate renewable energy readiness while downplaying transition challenges. This bias creates parallel information universes where people with different starting positions drift further apart despite access to the same underlying facts.

Optimism bias: We systematically overestimate our chances of experiencing positive events and underestimate negative ones. Applied to climate change, this manifests as "it won't be that bad" or "technology will save us without requiring difficult changes." It's like how most drivers consider themselves above average---a statistical impossibility that reveals our self-serving assumptions.

Tribalism: We judge information based on who communicates it rather than its inherent accuracy. The exact same climate facts presented by different messengers receive dramatically different credibility assessments based on perceived group membership. It's why climate scientists presenting identical information get wildly divergent reactions from different audiences based on perceived political alignment (Hornsey, 2021).

These biases don't make climate action impossible, but they do create predictable obstacles that effective engagement strategies must address. Simply providing more information without accounting for these cognitive tendencies rarely changes minds or behaviors. It's like trying to fill a bucket with a hole in the bottom---the information leaks out through these biases before it can accumulate into changed perspectives.

The good news is that awareness of these biases helps counteract them. Understanding why your brain naturally resists climate realities and solutions creates space for more deliberate, reflective processing. It's like having a map of the cognitive potholes on the road to climate action--- you can't repave the entire street, but you can at least navigate around the worst hazards.

Values and Worldviews: Why Facts Aren't Enough

If you've ever presented someone with rock-solid climate facts only to watch them remain completely unmoved, you've witnessed an important psychological truth: facts alone rarely change minds. This isn't because people are stupid or irrational (well, not entirely), but because we filter information through pre-existing values and worldviews that determine which facts get accepted and which get rejected.

Climate communication often assumes an "information deficit model"---if people just understood the science better, they'd naturally support climate action. Decades of evidence show this model is fundamentally flawed. On politically charged topics like climate change, more education and information often widen rather than narrow polarization, as people become better at finding evidence that supports their existing views (Hornsey, 2021).

Instead, our climate attitudes are shaped by deeper factors:

Cultural worldviews: Research on "cultural cognition" identifies two key dimensions that influence climate attitudes: hierarchy vs. egalitarianism, and individualism vs. communitarianism. Those with hierarchical and individualistic worldviews typically show more skepticism toward climate science, partly because traditional climate narratives can threaten their values by implying the need for collective action and regulatory constraints. It's not that these individuals can't understand climate science---it's that accepting it seems to undermine their deeper beliefs about how society should function.

Political identity: In some countries, climate attitudes have become tribal signifiers of political identity. Under these conditions, expressing the "wrong" climate views can risk social rejection from one's political community. It's like wearing a rival team's jersey to a sports bar---technically allowed, but likely to cause social friction. This identity-protective psychology explains why some of the most scientifically literate individuals can be the most polarized on climate change---they're better at rationalizing positions that align with their political tribe.

Personal experience: Lived experience with extreme weather, changing seasons, or other tangible climate impacts can sometimes bypass ideological filters, making climate change concrete rather than abstract. A lifetime conservative farmer who observes shifting growing seasons may accept climate realities that conflict with their political tribe's official position. Experience speaks in a language that can sometimes transcend political boundaries.

Economic interests: Perceived economic threats or opportunities significantly shape climate attitudes. Those who see their livelihoods threatened by climate policies (like fossil fuel workers) naturally resist them, while those who stand to benefit from a clean energy transition typically show greater support. These economic calculations often operate beneath conscious awareness, influencing which information seems credible and which doesn't (IPCC, 2022).

Moral foundations: Research suggests that different moral intuitions underlie political differences. Conservative-leaning individuals tend to value purity, authority, and loyalty alongside care and fairness, while liberal-leaning individuals emphasize care and fairness more predominantly. Traditional environmental messaging often frames climate change exclusively in terms of harm prevention (care) and justice (fairness), missing moral dimensions that resonate with more conservative audiences.

Understanding these deeper influences reveals why simply correcting misunderstandings or presenting additional scientific evidence rarely changes climate attitudes. It's like trying to fix a computer software problem by replacing the monitor---you're addressing the wrong level of the system.

Effective climate engagement requires connecting with these underlying values and worldviews, not just correcting factual misunderstandings.

This doesn't mean abandoning scientific accuracy, but rather recognizing that the same accurate information can be framed in ways that either threaten or affirm people's deeper identities and values.

From Paralysis to Action: Psychological Barriers and Bridges

Even for those who accept climate science and care deeply about the issue, psychological barriers can still prevent effective engagement. Understanding these obstacles---and how to overcome them---helps explain why climate concern doesn't automatically translate into climate action.

Barrier: Psychological distance Climate impacts often feel distant in time (affecting future generations), space (happening elsewhere), socially (affecting others unlike ourselves), and hypothetically (might not happen). This psychological distance reduces emotional engagement and urgency.

Bridge: Make it near Connect climate change to local, visible impacts. Discuss changes happening now, not just future projections. Emphasize effects on identifiable people and places, not just abstract statistics. Research shows that psychologically "near" threats generate stronger emotional responses and motivation for action (Marshall et al., 2021).

Barrier: Overwhelming scope The sheer magnitude of climate change can trigger feelings of helplessness and despair. If the problem seems insurmountably large, individual actions appear futile, leading to disengagement or apathy.

Bridge: Rightsize the response Break down overwhelming challenges into manageable pieces. Clarify how specific actions contribute to collective solutions. Connect individual behavior to community and policy engagement for greater impact. The key is matching the perceived scope of solutions to the perceived scope of the problem (NOAA, 2024).

Barrier: Apocalyptic framing Doom-filled climate narratives can paralyze rather than motivate. When climate futures seem hopeless, people often respond with denial, fatalism, or hedonism rather than constructive engagement.

Bridge: Empower with honest hope Balance clear-eyed recognition of climate risks with authentic hope based on existing solutions and ongoing progress. Research shows that efficacy---believing your actions matter---is essential for sustained engagement. Combine threat information with specific, accessible actions for the most effective response (Pihkala, 2022).

Barrier: Delayed feedback Climate actions rarely produce immediate, visible results. This delayed feedback makes it difficult to maintain motivation, like trying to lose weight on a scale that only shows changes years after diet modifications.

Bridge: Create feedback loops Develop proximate goals with observable outcomes. Celebrate progress through emissions reductions, community milestones, and policy achievements. Connect abstract carbon reductions to tangible co-benefits like cleaner air, cost savings, or community resilience.

Barrier: Social norms Humans are profoundly influenced by what others around them do and value. When climate-friendly behaviors seem unusual or socially costly, adoption remains limited despite personal concern.

Bridge: Leverage social influence Highlight the growing normality of climate action rather than its exceptional nature. Strengthen social rewards for climate-friendly choices through community recognition and shared identity. Research consistently shows that "what others like me are doing" strongly influences behavior, often more than environmental concern alone (Clayton et al., 2023).

Barrier: Sacrificial framing Climate action often gets framed as sacrifice, deprivation, and loss---giving up comforts and conveniences for abstract future benefits. This framing activates loss aversion, a powerful psychological force against change.

Bridge: Emphasize co-benefits and gain frames Highlight immediate benefits of climate-friendly choices---health improvements from active transportation, cost savings from energy efficiency, community connections from local engagement. Research shows that gain frames motivate action more effectively than loss frames for preventive behaviors.

These psychological bridges don't magically eliminate all barriers to climate action, but they do create more favorable conditions for engagement. Just as climate solutions must address both technological and economic challenges, they must also accommodate psychological realities. The most brilliant climate policies and technologies will fail if they don't account for how real humans actually think, feel, and make decisions.

Climate Communication: It's Not What You Say, It's How You Say It

If you've ever tried to discuss climate change with someone holding different views, you know that the same information can land very differently depending on how it's communicated. Climate communication isn't just about accurate content---it's about effectively engaging diverse audiences with different values, concerns, and information-processing styles.

Here are some evidence-based principles for more effective climate communication:

Know your audience: Different groups require different approaches. A message that resonates with environmental activists might alienate rural conservatives. A presentation compelling to scientists might bore business leaders. Effective communication starts with understanding your specific audience's values, concerns, and preferred communication styles. There's no one-size-fits-all approach to climate communication.

Find common values: Connect climate issues to values your audience already holds---whether that's conserving natural heritage, protecting children's health, advancing innovation, promoting national security, or being good stewards of creation. Research shows that value-congruent framing dramatically increases message acceptance. It's like finding the particular doorway through which climate information can enter someone's existing belief structure (Yale Program on Climate Change Communication, 2021).

Use trusted messengers: People evaluate information based partly on the perceived trustworthiness and similarity of the communicator. Military leaders can reach veterans and security-minded audiences. Faith leaders can engage religious communities. Local business owners can influence other entrepreneurs. The same message delivered by different messengers produces dramatically different results based on perceived shared values and identity.

Tell better stories: Humans are narrative creatures---we process information better through stories than through abstract data. Effective climate communication often uses narrative structures with identifiable characters, challenges, and resolutions. This doesn't mean abandoning science, but rather embedding scientific insights within stories that engage human emotion and imagination (Climate-XChange, 2020).

Use visual communication: Visual information processing often bypasses some of the cognitive filters that screen out inconvenient textual or verbal information. Well-designed graphics, maps, and images can convey climate impacts and solutions more effectively than words alone. The old adage that "a picture is worth a thousand words" holds particular truth for climate communication.

Focus on solutions: Research consistently shows that solution-focused climate communication generates more engagement and less defensive responding than exclusively threat-focused messaging. Without accessible solutions, threat information often produces fear and avoidance rather than constructive action. The ideal balance typically combines clear problem acknowledgment with proportionate, achievable solutions.

Leverage social norms: Highlighting what others are doing---especially others similar to your target audience---powerfully shapes behavior. Messages like "Most households in your neighborhood have already installed insulation" typically outperform purely informational or environmental appeals. Humans are social creatures who look to others for behavioral guidance, often unconsciously.

Use concrete, experiential language: Abstract climate concepts become more engaging when translated into concrete, sensory language. Instead of "2°C warming by mid-century," try "summers like the record-breaking 2023 heat wave becoming typical within 30 years." Research shows that concrete language activates more brain regions and creates stronger memory and motivation than abstract concepts.

These principles don't guarantee success in every climate conversation, but they substantially improve the odds of meaningful engagement. They address the reality that human information processing involves emotion, identity, and social context---not just logical evaluation of facts. The most scientifically accurate climate message still fails if it never manages to enter and influence someone's thinking.

This science-based approach to communication isn't about manipulating people---it's about respecting how human cognition actually works rather than how we might wish it worked. Just as we wouldn't try to run a gasoline engine on water because the physics doesn't work, we shouldn't expect climate communication to succeed without accounting for the psychology of how information influences beliefs and behaviors.

Climate Anxiety: When Awareness Becomes Overwhelming

As climate impacts intensify and receive greater media coverage, more people experience climate anxiety---distress related to awareness of the climate crisis and concern about its implications. This emotional response ranges from mild worry to clinical anxiety affecting daily functioning. It's particularly prevalent among young people, with studies showing significant percentages experiencing climate distress that interferes with normal activities (IPCC, 2022).

Climate anxiety isn't necessarily pathological---it's often a rational response to a genuine threat. The problem occurs when these emotions become so overwhelming that they prevent constructive engagement rather than motivating it. It's like how moderate stress can enhance performance, but excessive stress becomes paralyzing.

Several factors can transform normal climate concern into debilitating anxiety:

Media immersion: Constant exposure to catastrophic climate narratives without contextual solutions or progress updates can create apocalyptic thinking. The 24/7 disaster-focused news cycle rarely presents balanced perspectives showing both challenges and responses. It's like watching an endless loop of healthcare crisis stories without ever seeing hospitals, doctors, or treatments.

Perceived helplessness: When climate threats seem overwhelming and effective responses seem inaccessible, anxiety flourishes. This perceived lack of agency---feeling unable to influence outcomes that matter deeply---creates particular psychological distress. It's the difference between being in a fast-moving car as a driver versus as a passenger with no access to the controls.

Social isolation: Processing climate concerns alone, without supportive communities acknowledging both the emotional and practical dimensions of the crisis, intensifies anxiety. Humans are social creatures who manage threats better collectively than individually. Climate anxiety often peaks when people feel they're facing an existential crisis without social support.

Temporal distance: The worst projected climate impacts often appear in timeframes that current young people will live to experience, while decision-making power remains concentrated in older generations with shorter future time horizons. This creates a particular form of generational climate anxiety---concern about inheriting problems created by others with limited power to address them.

Fortunately, research and clinical experience suggest several approaches for managing climate anxiety:

Validate emotions: Acknowledging that climate anxiety reflects appropriate concern for a real problem, not irrational fear, creates space for constructive engagement. Suppressing or dismissing these emotions ("don't worry about things you can't control") typically backfires, driving anxiety underground rather than resolving it.

Build agency through action: Meaningful engagement with climate solutions---whether personal lifestyle changes, community initiatives, or policy advocacy---provides antidotes to helplessness. Research consistently shows that feeling effective is crucial for emotional wellbeing when facing large challenges.

Create supportive communities: Climate cafés, action groups, faith communities, and other settings where climate emotions can be processed collectively help transform isolation into connection. Sharing concerns and responses with others provides both emotional support and practical pathways forward.

Practice climate-aware self-care: Maintaining physical and emotional wellbeing through adequate rest, time in nature, creative expression, and meaningful connection builds resilience against anxiety. Self-care isn't selfish or distraction---it's necessary maintenance for sustained climate engagement.

Cultivate realistic hope: Neither blind optimism nor fatalistic pessimism serves wellbeing or effective action. Realistic hope acknowledges serious challenges while recognizing genuine progress, available solutions, and capacity for collective action. It's grounded in evidence rather than wishful thinking or despair (Ojala, 2021).

The goal isn't eliminating all climate distress---some emotional response to a genuine crisis is appropriate and motivating. Rather, the aim is transforming debilitating anxiety into constructive concern that fuels effective engagement. This balanced approach recognizes both the emotional and practical dimensions of climate response. After all, we need people engaged for the long haul, not burned out by unsustainable levels of distress.

Identity and Meaning: Climate in the Story of Your Life

Beyond specific psychological barriers and communication techniques, climate engagement connects to deeper questions of identity and meaning. How does climate change fit into your understanding of who you are and what gives your life purpose? These existential dimensions might seem abstract compared to carbon budgets and renewable energy technologies, but they profoundly influence how we respond to the climate crisis.

Climate change challenges several aspects of identity and meaning:

Future orientation: Climate change confronts us with questions about what kind of future we're creating and what legacy we'll leave. This challenges us to extend our moral consideration beyond immediate concerns to future generations and distant others. It asks whether we're living in ways aligned with the future we wish to create (Clayton et al., 2023).

Relationship with nature: The climate crisis reveals the consequences of treating nature as merely resources to exploit rather than complex systems upon which we depend. This challenges dominant cultural narratives about human separateness and superiority, inviting reconsideration of our place within rather than above natural systems.

Consumption and status: Climate solutions often involve shifting from high-consumption lifestyles to more sustainable alternatives. This challenges consumer identities and status markers built around material acquisition and carbon-intensive displays like large homes, vehicles, and frequent flying (IPCC, 2022).

Collective vs. individual: Climate change is inherently a collective problem requiring coordinated responses across society. This challenges hyper-individualistic frameworks that locate both problems and solutions primarily at the individual level. It asks how we balance personal freedom with collective responsibility.

These identity and meaning challenges help explain why climate change triggers such strong responses, both negative and positive. For some, climate awareness creates existential threat, undermining cherished beliefs about progress, nature, consumption, and individualism. For others, it offers opportunities for growth and purpose---chances to align actions more closely with deeper values and to participate in historically significant transformation.

Rather than threats to identity, climate engagement can offer several sources of meaning and purpose:

Legacy: Contributing to climate solutions connects individual actions to something larger and more enduring than personal comfort or achievement. It creates legacy by helping preserve livable conditions for future generations. This taps into deep human desires to matter and make positive differences that outlast our individual lives.

Community: Climate action inherently involves working with others toward shared goals. This creates meaning through connection, belonging, and shared purpose---fundamental human needs often underserved in individualistic consumer societies. Many climate activists report that the relationships formed through collective engagement provide as much sustenance as the cause itself.

Authenticity: Aligning daily choices with climate awareness helps reduce cognitive dissonance between what we know and how we live. This integration of knowledge, values, and behavior creates authenticity and integrity---the sense that our outer lives reflect our inner convictions rather than contradicting them.

Growth: Engaging with climate challenges us to develop new skills, perspectives, and capacities. This growth---whether learning about energy systems, practicing effective communication, building community organizing skills, or developing emotional resilience---creates meaning through continuous development and mastery.

These meaning sources don't eliminate the difficulties of climate engagement, but they provide sustaining resources for the journey. They transform climate action from mere obligation or sacrifice into an integral part of a purposeful, values-aligned life. Rather than threatening identity, climate engagement can strengthen it by connecting personal choices to enduring values and larger purposes.

This meaning-centered approach doesn't replace practical climate solutions---we still need renewable energy, electrified transportation, sustainable agriculture, and climate policy. But it provides the psychological and spiritual resources to pursue these solutions with sustained commitment rather than burnout or disillusionment. The climate challenge isn't just technical and political---it's also existential and moral, requiring resources of meaning and purpose alongside technology and policy.

The Psychology of Hope: Finding the Sweet Spot Between Denial and Despair

Hope might seem like a minor factor compared to emissions targets and technology deployment, but psychological research increasingly recognizes its crucial role in climate engagement. Without realistic hope, climate awareness often leads to either denial (psychologically rejecting threatening information) or despair (accepting the threat but rejecting possibility of adequate response). Both reactions prevent effective engagement (Ojala, 2021).

The psychological dynamics around climate hope are complex:

Too little hope leads to fatalism, disengagement, and present-focused hedonism ("why bother if we're doomed"). Without perceived response efficacy---belief that effective solutions exist---awareness of climate threats often produces paralysis rather than action. It's like telling someone they have a serious illness without mentioning available treatments---the information just produces despair.

False hope based on magical thinking or technology salvation fantasies creates complacency and delays necessary transitions. Assuming future technologies will effortlessly solve climate challenges without present action or that minimal changes will suffice despite contrary evidence undermines appropriate urgency. It's like expecting to ace an exam without studying because something will magically work out.

Realistic hope acknowledges both the severity of climate threats and genuine progress alongside available solutions. This balanced perspective supports sustained engagement by connecting honest problem assessment with achievable response pathways. It's like acknowledging a serious medical condition while recognizing effective treatments exist if promptly applied.

Cultivating realistic climate hope involves several components:

Evidence-based progress assessment: Tracking genuine climate progress alongside continuing challenges provides reality checks against both unwarranted pessimism and complacency. The rapid growth of renewable energy, declining coal use, increasing policy ambition, and expanding climate movements all provide empirical grounds for hope without minimizing remaining challenges.

Possibility-focused thinking: Distinguishing between what's guaranteed (positive or negative) and what's possible expands perceived agency. While neither climate catastrophe nor complete resolution is guaranteed, multiple futures remain possible depending on choices and actions we take now. This possibility space creates room for meaningful engagement rather than predetermined outcomes.

Complex systems understanding: Recognizing that climate and social systems contain nonlinear dynamics, tipping points, and emergent properties creates space for transformative rather than merely incremental change. Small interventions at key leverage points can potentially catalyze systemic shifts beyond linear projections of current trends. This systems thinking counters both naive optimism and reflexive pessimism by highlighting contingency and multiple pathway possibilities.

Historical perspective: Examining previous societal transformations--- from civil rights to smoking reduction to renewable energy growth--- demonstrates that significant social and technological change can happen faster than linear projections suggest once key thresholds are crossed. These historical examples don't guarantee similar climate progress but demonstrate that rapid change isn't unprecedented when conditions align (NASA, 2024).

Multiple scales of agency: Finding appropriate scope for personal agency prevents both inflated expectations of individual impact (leading to burnout when reality disappoints) and complete dismissal of individual contribution (leading to helplessness). Engaging at personal, community, and advocacy levels creates multilevel agency matched to the multilevel challenge, supporting sustainable hope through realistic contribution.

These approaches create psychological conditions for what scholars call "active hope"---not optimism about guaranteed outcomes but commitment to meaningful possibilities despite uncertainty. This active hope differs fundamentally from both passive optimism (everything will work out regardless of action) and resigned pessimism (nothing will work regardless of action). It's hope as practice rather than prediction, assessed by ongoing engagement rather than emotional state alone.

The ability to maintain realistic hope may be among the most important psychological resources for effective climate engagement. Without it, even the most knowledgeable individuals tend toward either denial or despair when confronting climate realities. With it, even daunting challenges become approachable through sustained, strategic engagement focused on meaningful possibility rather than guaranteed outcomes.

Conclusion: Changing Minds, Changing Climate

Throughout this chapter, we've explored the complex psychology of climate change---from cognitive biases and cultural worldviews to communication strategies and hope cultivation. This psychological terrain helps explain why climate progress often falls short of what scientific understanding would suggest is necessary and possible. It's not that humans are incapable of understanding climate science or responding effectively to the challenge---it's that psychological and cultural factors mediate between knowledge and action in ways that often delay, dilute, or derail effective response (Clayton et al., 2023).

Yet this same psychological understanding offers pathways for more effective engagement. By designing climate communications, policies, and action strategies that work with rather than against human psychology, we can accelerate progress and avoid common pitfalls. This doesn't mean manipulating people---it means respecting how real minds work rather than assuming humans are perfectly rational processors of abstract information.

Several insights stand out as particularly valuable for enhanced climate engagement:

Connect with values, not just facts: Effective climate engagement connects with diverse audiences through their existing values rather than assuming shared environmental concerns. Whether economic opportunity, national security, public health, innovation leadership, or moral responsibility, multiple pathways exist to climate concern and action beyond purely environmental framing.

Build agency at multiple scales: Combining personal, community, and policy engagement creates multilevel agency matched to the multilevel climate challenge. This comprehensive approach prevents both the limitations of purely individual action and the helplessness of believing only system-level change matters. It creates realistic pathways for contribution regardless of position or circumstance.

Foster post-traumatic growth: Climate engagement offers potential for growth through meaning creation, expanded identity, deepened relationships, and new capabilities. This growth perspective transforms climate action from mere obligation or sacrifice into pathway for personal and collective development. It creates psychological sustainability for the long journey of climate engagement (Pihkala, 2022).

Leverage social influence: Humans are profoundly social creatures whose behavior is shaped by perceived norms and group identities. Creating visible examples of climate-friendly practices, strengthening climate-concerned identities, and normalizing sustainable behaviors harnesses these social dynamics for positive change. Individual behavior and structural transformation both emerge from and contribute to evolving social norms.

Cultivate realistic hope: Balancing honest assessment of climate threats with recognition of meaningful progress and available solutions creates psychological conditions for sustained engagement. This realistic hope differs from both naive optimism and fatalistic pessimism, supporting active response rather than either complacency or paralysis.

These psychological insights don't replace the need for scientific understanding, technological innovation, policy reform, and economic transformation. Rather, they provide connective tissue between these domains and actual human beings whose beliefs, values, emotions, and actions ultimately determine whether and how quickly these larger changes occur.

The climate challenge isn't just "out there" in atmospheric composition, energy systems, and political institutions---it's also "in here" in human minds struggling to comprehend, process, and respond to an unprecedented global threat. Addressing the climate crisis effectively requires engagement with both external systems and internal psychological realities.

The good news is that humans possess remarkable capacity for psychological growth, cultural evolution, and collective action when properly supported. The same cognitive flexibility, social learning, and moral imagination that created our current predicament can also help navigate toward more sustainable futures. The climate is changing---and with the right psychological understanding and support, we can change too, hopefully fast enough to preserve a livable planet for ourselves and future generations.

In the next chapter, we'll examine climate adaptation---the necessary complement to emissions reduction as we learn to live with the climate changes already underway and those still to come despite our best mitigation efforts. Because even with the most ambitious climate action, some warming remains unavoidable, requiring proactive adaptation alongside transformative mitigation.

Chapter 13: Adapting Without Giving Up

When Prevention Isn't Enough

Remember when your parents told you to wear a helmet while biking? That's risk prevention. Now remember when they also taught you how to bandage a scraped knee? That's adaptation—preparing for the reality that sometimes prevention isn't enough. In climate terms, we've spent most of this book discussing the climate equivalent of helmets—reducing emissions to prevent the worst impacts. But it's time to talk about bandages too, because even with our most ambitious mitigation efforts, some climate impacts are already locked in.

Let's be blunt: the climate is changing, and will continue changing for decades even under the most optimistic emissions scenarios. We've already warmed about 1.2°C above pre-industrial levels, and we're seeing the consequences: stronger storms, longer droughts, rising seas, shifting seasons, and more extreme heat. Even if we magically stopped all greenhouse gas emissions tomorrow (spoiler alert: we won't), the climate would continue changing due to the greenhouse gases already in the atmosphere. It's like how a pot of water keeps boiling for a while after you turn off the stove—there's thermal inertia in the system (IPCC, 2022).

This reality forces a difficult but necessary conversation about adaptation—the process of adjusting to actual or expected climate change and its effects. Adaptation isn't about giving up on mitigation or accepting a hopelessly damaged climate future. It's about recognizing that some climate change is now unavoidable and preparing to protect communities, ecosystems, and infrastructure from its worst impacts while we simultaneously work to reduce emissions.

In this chapter, we'll explore climate adaptation across sectors—from infrastructure and agriculture to public health and ecosystem management. We'll examine successful adaptation strategies, acknowledge their limitations, and navigate the thorny ethical questions adaptation raises. And throughout, we'll maintain the delicate balance between acknowledging unavoidable impacts and avoiding fatalistic resignation. Because the most dangerous response to climate change isn't either mitigation or adaptation—it's the paralysis that comes from believing it's too late to do anything at all.

The Adaptation Spectrum: From Coping to Transformation

Before diving into specific adaptation approaches, it helps to understand the spectrum of adaptation strategies. Not all adaptation looks the same —it ranges from simple coping mechanisms to complete systemic transformation, with important differences in ambition, cost, and effectiveness (Pelling et al., 2015).

Coping responses are reactive, short-term measures deployed after climate impacts occur. Think of emergency evacuations during floods, temporary water restrictions during droughts, or setting up cooling centers during heat waves. These approaches resemble treating symptoms rather than addressing underlying vulnerabilities. They're necessary but insufficient—like using pain relievers for a chronic condition without addressing its root cause.

Incremental adaptation involves modest adjustments to existing systems without fundamentally changing their structure. Examples include building slightly higher seawalls for coastal protection, developing more drought-resistant crop varieties, or upgrading stormwater drainage systems for heavier rainfall. These approaches acknowledge climate risks but attempt to preserve current systems and practices with minimal disruption. It's like upgrading your umbrella for heavier rain rather than questioning whether you should be walking in the storm at all.

Transformational adaptation involves fundamental changes to systems in response to climate impacts that overwhelm incremental approaches. This might mean relocating entire communities from areas that can no longer be protected from sea level rise, switching to entirely different agricultural systems better suited to new climate conditions, or redesigning cities for dramatically different temperature and precipitation patterns. These approaches recognize that in some cases, preserving the status quo is neither possible nor desirable. It's like deciding to move to a different climate altogether rather than buying a bigger air conditioner each year.

As climate impacts intensify, we're discovering that coping and incremental approaches—while politically easier—often prove insufficient in the face of accelerating changes. The floods, fires, droughts, and heat waves we're already experiencing are pushing many systems beyond their coping capacity, forcing consideration of more transformational approaches. It's like discovering that your trusty umbrella, even with upgrades, simply can't handle the new reality of your region's monsoon-like downpours.

The most robust adaptation strategies combine elements across this spectrum—employing coping mechanisms for immediate threats while developing incremental adaptations for the medium term and planning transformational approaches for the longer term. This multi-layered approach creates both immediate protection and pathways to more fundamental changes as conditions evolve.

But here's where it gets tricky: how do we know which level of adaptation is appropriate in a given situation? When is incremental change sufficient, and when do we need transformation? These aren't just technical questions—they involve values, risk tolerance, resource constraints, and ethical considerations about who decides and who benefits. They require balancing near-term protection with long-term sustainability in contexts of significant uncertainty—challenges that are as much about governance, equity, and culture as they are about climate science and engineering.

Infrastructure Adaptation: Building for the New Normal

Our built environment—roads, bridges, buildings, power plants, water systems—was designed for the climate of the past. Most infrastructure was engineered using historical weather data that no longer reflects current conditions, let alone future extremes. It's like trying to drive using only your rearview mirror—you're navigating based on where you've been, not where you're going (ASCE, 2021).

This mismatch between design parameters and emerging climate realities creates vulnerabilities across all infrastructure sectors:

Transportation systems face threats from extreme heat buckling railways and softening asphalt, flooding undermining roads and bridges, and sea level rise inundating coastal highways and airports. The 2021 heat dome in the Pacific Northwest provided a preview when temperatures reached 122°F on roadways, causing pavement to expand and crack. Meanwhile, increasingly frequent "100-year floods" regularly overtop bridges designed for historical flow patterns. These disruptions don't just inconvenience travelers—they block emergency response, interrupt supply chains, and isolate vulnerable communities.

Energy infrastructure faces multiple climate challenges, from reduced hydropower during droughts to transmission line failures during extreme heat to power plant cooling limitations as water temperatures rise. During the 2022 European heat wave, multiple nuclear plants had to reduce output because the river water used for cooling became too warm to effectively remove heat while meeting environmental regulations for discharge temperatures. Meanwhile, in Texas, unprecedented winter cold in 2021 caused catastrophic power failures when natural gas infrastructure froze—an adaptation failure with deadly consequences (Busby et al., 2021).

Water systems designed for historical precipitation patterns struggle with both too much water (overwhelming stormwater systems during intense rainfall) and too little (depleting reservoirs during extended droughts). Cities like Phoenix face the prospect of major water shortages as the Colorado River's flow diminishes, while places like Houston contend with drainage systems designed for rainfall patterns that no longer exist. These challenges threaten both water quantity and quality—as declining river flows concentrate pollutants and overwhelming storm events flush contaminants into water bodies (EPA, 2023).

Buildings designed for previous temperature ranges increasingly require retrofit for both extreme heat and, in some regions, unprecedented cold.

The deadly 2021 heat wave in the Pacific Northwest highlighted how housing built for historically moderate climates left residents dangerously vulnerable when temperatures soared beyond anything the region had experienced. Similar vulnerabilities appear in cold-weather extremes, as buildings designed for mild winters proved inadequate during unusual freezing events in typically warm regions.

Adapting infrastructure for climate resilience requires several complementary approaches:

Climate-informed design standards that incorporate forward-looking climate projections rather than historical data alone. This means designing bridges for future flood levels, not just past records; sizing stormwater systems for increasingly intense precipitation; and building power lines that can withstand stronger winds and higher temperatures. It's like buying clothes for a growing child—you need to account for future conditions, not just current fit.

Hardening critical infrastructure against climate extremes through physical modifications like elevated electrical equipment in flood zones, expanded culverts under roadways, increased capacity for water storage and drainage, and strengthened transmission towers. These approaches acknowledge that some infrastructure can't be easily relocated and must instead be reinforced against emerging threats.

Incorporating redundancy and flexibility to maintain function during disruptions through backup systems, alternative routes, microgrids, and modular designs that can be quickly repaired or modified. This doesn't just mean duplicate systems but genuinely diverse approaches that don't share the same vulnerabilities. It's the infrastructure equivalent of diversifying your investment portfolio—

spreading risk across different approaches rather than putting all your eggs in one basket.

Strategic retreat from areas where long-term protection isn't feasible, such as relocating infrastructure from chronically flooded zones or regions facing inevitable inundation from sea level rise. While politically challenging, strategic retreat often proves more cost-effective than endless cycles of damage and repair in highly vulnerable locations. It's like knowing when to fold in poker rather than throwing good money after bad.

Nature-based solutions that work with natural processes rather than against them, such as restoring wetlands for flood mitigation, expanding urban tree canopy for cooling, and protecting coastal ecosystems as buffers against storms and sea level rise. These approaches often provide multiple benefits beyond climate adaptation—enhancing biodiversity, improving air and water quality, creating recreational spaces, and sequestering carbon.

Perhaps the most challenging aspect of infrastructure adaptation is its timeframe. Infrastructure typically lasts decades, sometimes centuries. Decisions made today about where and how to build will shape vulnerability for generations. This long-term perspective clashes with political cycles focused on immediate concerns and visible results. Yet failing to incorporate climate projections into infrastructure decisions today guarantees greater costs and disruptions tomorrow—a classic case of prevention being far cheaper than cure.

Agricultural Adaptation: Feeding a Hotter World

Agriculture sits on the front lines of climate change—directly exposed to shifting temperatures, changing precipitation patterns, extreme weather events, and ecological disruptions. Farmers have always adapted to weather variability, but climate change is pushing many agricultural systems beyond their adaptive capacity with unprecedented combinations of heat, drought, flooding, and pest pressure (FAO, 2021).

The climate challenges to agriculture are multifaceted:

Temperature changes affect crop growth, livestock health, and labor conditions. Many staple crops have temperature thresholds above which yields decline dramatically—corn pollination fails above about 95°F, while wheat protein content decreases in extreme heat. Livestock face heat stress that reduces productivity and increases mortality. And farmworkers confront increasingly dangerous conditions as wet-bulb temperatures approach human physiological limits in many agricultural regions (Mora et al., 2017).

Water availability fluctuates more dramatically with intensifying droughts and floods. The agricultural equation has always balanced water supply and demand, but climate change is disrupting this balance through both long-term trends (like earlier snowmelt and reduced summer river flows) and short-term extremes (like flash droughts and atmospheric rivers). Regions built around irrigation increasingly face water allocation conflicts between agriculture, cities, industry, and ecological needs.

Pest, disease, and weed pressures shift with changing climatic conditions. Warming temperatures allow agricultural pests to expand their ranges, survive through previously limiting winter conditions, and complete more generations per season. Meanwhile, beneficial insects like pollinators face disruptions to their life cycles and habitat availability. The timing mismatches between crops, pests, and beneficial organisms create novel management challenges that existing practices often cannot address.

Extreme weather events like hailstorms, flash floods, and high winds cause direct crop damage and soil erosion. Agricultural systems optimized for historical climate conditions often lack resilience to these intensifying extremes. A single ill-timed storm can destroy an entire season's production, while sequential extremes can degrade soil health and agricultural infrastructure over longer periods.

Agricultural adaptation employs multiple strategies across timeframes:

Short-term tactical adjustments include changing planting dates to avoid seasonal heat stress, switching to more heat- or drought-tolerant varieties, modifying irrigation scheduling, and adapting pest management practices to new pressures. These approaches work within existing agricultural systems while adjusting specific practices to emerging conditions. They're like changing your route to work to avoid new traffic patterns—still driving the same car to the same destination, just navigating differently.

Medium-term strategic shifts involve more substantial changes like adopting new crop rotations, investing in water-efficient irrigation systems, implementing cover cropping and reduced tillage for soil health, and diversifying production to spread risk. These approaches modify agricultural systems to enhance resilience while maintaining their basic structure. They're like remodeling your house for better energy efficiency—significant investment that improves performance without complete reconstruction.

Long-term transformational approaches include transitioning to entirely different production systems better suited to emerging climate conditions, such as agroforestry in regions facing increased heat and drought, indoor controlled-environment agriculture where field production becomes unviable, or completely new crop and livestock combinations aligned with shifting climatic zones. These approaches acknowledge that maintaining historical agricultural systems may prove impossible in some regions despite incremental adaptations. They're like moving to an entirely different climate when your current location becomes unlivable—a profound shift rather than marginal adjustment.

Successful agricultural adaptation often combines three key elements:

Climate-resilient genetics through both traditional breeding and newer techniques that develop crops and livestock with improved tolerance to heat, drought, flooding, and emerging pest pressures. These efforts expand the genetic toolkit available to farmers facing changing conditions. While controversial in some contexts, genetic approaches (both traditional and modern) provide crucial adaptation pathways, especially when combined with appropriate management practices.

Regenerative management practices that build soil health, enhance biodiversity, improve water retention, and increase overall system resilience. Approaches like cover cropping, reduced tillage, adaptive grazing, and diversified rotations create agricultural systems better able to withstand climate stresses while often sequestering carbon and reducing emissions. These practices represent "no regrets" adaptations that deliver benefits regardless of specific climate trajectories (FAO, 2023).

Knowledge systems and decision support that help farmers navigate increasing uncertainty through improved forecasting, early warning systems, climate-informed planning tools, and peer learning networks. The accelerating pace of change means historical experience provides less reliable guidance for agricultural decisions, increasing the value of forward-looking information and collective knowledge sharing.

Throughout agricultural adaptation runs a central tension: balancing climate resilience with productivity, profitability, and food security. Adaptation strategies often require upfront investment before delivering long-term benefits, creating difficult tradeoffs for farmers operating on thin margins. Meanwhile, agricultural policies and market structures frequently incentivize short-term production at the expense of long-term resilience. Resolving these tensions requires aligning economic signals with climate realities through reformed subsidies, ecosystem service payments, risk management tools, and market structures that reward resilient practices.

Public Health Adaptation: Protecting People in a Changing Climate

When we think about climate impacts, dramatic visuals like hurricanes and wildfires usually come to mind. But some of the most significant climate threats operate more subtly through human health. From heat-related illnesses and vector-borne diseases to air quality degradation and mental health impacts, climate change affects health through multiple pathways—many of which remain underappreciated in climate discussions.

The health impacts of climate change include both direct and indirect effects:

Extreme heat directly threatens human physiology, with particular risk to older adults, young children, outdoor workers, people with pre-existing conditions, and those lacking access to cooling. Heat-related deaths often go underreported because heat exacerbates existing conditions rather than appearing as the primary cause of death. Yet studies consistently show excess mortality during heat waves—like the estimated 70,000 additional deaths during Europe's 2003 heat wave. As warming continues, many regions face the prospect of heat beyond human adaptive capacity, where even healthy individuals cannot survive prolonged exposure (Ebi et al., 2021).

Changing disease patterns emerge as warming temperatures allow disease vectors like mosquitoes and ticks to expand their ranges and remain active for longer seasons. Diseases previously limited to tropical regions increasingly threaten temperate zones, while formerly controlled diseases reemerge in new areas. For example, dengue fever—a debilitating viral infection transmitted by mosquitoes—has expanded its global range by about 30% in recent decades due partly to climate factors. Similarly, tick-borne diseases like Lyme disease have spread into regions where they were previously rare or absent.

Air quality degradation occurs through multiple climate-related pathways: increased pollen production and longer allergy seasons; more frequent and intense wildfires producing dangerous smoke; higher temperatures accelerating ground-level ozone formation; and changing atmospheric conditions that trap pollutants near population centers. These air quality impacts exacerbate respiratory and cardiovascular diseases like asthma, COPD, and heart disease, with particular risk to vulnerable populations (WHO, 2021).

Mental health impacts range from trauma and displacement following extreme weather events to chronic stress, anxiety, and grief associated with both experienced and anticipated climate changes. Climate-related disasters can trigger post-traumatic stress disorder, depression, and substance abuse, while longer-term environmental changes contribute to conditions now recognized as solastalgia—distress caused by environmental change affecting one's home environment. These mental health dimensions of climate change remain understudied but increasingly recognized as significant public health concerns (Clayton et al., 2017).

Food and water insecurity intensifies as climate change disrupts agricultural systems, contaminates water supplies during flooding events, and exacerbates both chronic and acute malnutrition. These impacts operate through complex pathways, from direct production losses to price spikes affecting food access to infrastructure damage disrupting distribution systems. The health consequences cascade through compromised immune function, developmental impacts on children, increased vulnerability to disease, and exacerbated social tensions in resource-stressed communities.

Public health adaptation employs multiple strategies to address these diverse impacts:

Heat action plans combine early warning systems, cooling centers, targeted outreach to vulnerable populations, adjusted work schedules for outdoor labor, and enhanced emergency medical response during extreme heat events. These plans increasingly integrate elements like cool roofs, expanded tree canopy, cooling pavements, and green infrastructure to reduce urban heat island effects. The most effective approaches address both immediate heat response and longer-term urban cooling through design and policy changes.

Disease surveillance and vector control systems monitor changing disease patterns and implement preventive measures from mosquito control to vaccination campaigns to public education about disease risks. These systems require increased capacity and geographic coverage as climate change extends the range and season of various disease vectors. They also benefit from climate-informed forecasting that anticipates changing risk patterns based on seasonal and long-term climate projections.

Air quality management addresses both acute episodes (like wildfire smoke advisories with indoor air filtration recommendations) and chronic conditions through emissions regulations, transportation planning, energy policy, and land use decisions. Climate-informed air quality management recognizes the interactions between climate factors and pollution sources, developing integrated approaches rather than treating these as separate issues.

Mental health support includes both crisis response following climate disasters and longer-term services addressing ongoing climate distress. Psychological first aid during evacuations, displacement, and recovery has become an increasingly important component of disaster response. Meanwhile, therapeutic approaches specifically addressing climate anxiety and grief continue developing as these conditions become more widespread. Community-based mental health promotion through social connection, meaningful engagement, and collective action provides additional protective factors against climate-related psychological distress.

Food and water safety systems monitor and respond to climate-related contamination risks, supply disruptions, and nutritional challenges through enhanced surveillance, emergency response capacity, and preventive infrastructure investments. These systems increasingly incorporate climate projections into planning for issues like harmful algal blooms in drinking water sources, bacterial contamination during flooding events, and nutritional security during agricultural disruptions.

Public health adaptation faces several distinctive challenges:

Social determinants of health shape vulnerability to climate impacts, with socioeconomic disparities, structural racism, and access barriers creating disproportionate risks for marginalized communities. Effective adaptation requires addressing these underlying inequities rather than treating only proximate health impacts. This means integrating climate health responses with broader social justice efforts addressing income inequality, environmental injustice, and health care access (USGCRP, 2018).

Reactive health systems designed for known, historical health challenges struggle to anticipate and prevent emerging climate-related health threats. Shifting toward preventive, adaptive approaches requires fundamental changes in health system design, funding priorities, workforce training, and intersectoral collaboration. Climate-informed health systems must balance immediate health needs with longer-term preparedness in contexts of limited resources and competing priorities.

Cross-sectoral dependencies mean that public health adaptation cannot succeed in isolation from adaptations in energy, water, food, and infrastructure systems. Protecting health during heat waves, for example, requires reliable electricity for cooling, water systems for hydration, functional transportation for emergency response, and food systems that continue operating during extreme conditions. These interdependencies necessitate integrated adaptation planning across traditionally siloed sectors.

Despite these challenges, public health adaptation offers particularly favorable cost-benefit ratios compared to many other adaptation domains. Preventing heat-related hospitalizations costs far less than treating them. Early vector control prevents disease outbreaks cheaper than treating epidemics. Mental health support reduces both human suffering and productivity losses at a fraction of their economic cost. These preventive approaches represent "no regrets" adaptations that provide benefits under any climate scenario while improving wellbeing regardless of specific climate trajectories.

Ecosystem Adaptation: Helping Nature Navigate Climate Change

Humans aren't the only species facing climate adaptation challenges. Ecosystems worldwide confront unprecedented rates of change that threaten biodiversity, ecosystem services, and ecological resilience. While ecosystems have adapted to climate changes throughout Earth's history, current changes occur at rates 10-100 times faster than most natural climate shifts, outpacing many species' adaptive capacity (IPBES, 2019).

Ecosystems face multiple interacting climate stressors:

Temperature changes affect species' physiology, behavior, and geographic distributions. Many species have specific temperature tolerances beyond which they cannot survive, reproduce, or compete effectively. As warming pushes beyond these thresholds, species either adapt (which may take many generations), migrate to more suitable areas (if available and accessible), or face local extinction. Species unable to adapt or migrate quickly enough—due to limited dispersal ability, habitat fragmentation, or specialized requirements—face particularly high extinction risks.

Precipitation shifts alter water availability, timing, and intensity, affecting everything from plant growth patterns to aquatic habitats. Regions experiencing increased drought stress see shifts from mesic to xeric vegetation communities, while areas with intensifying precipitation face soil erosion, landslides, and altered river morphology. These hydrological changes ripple through ecosystems, affecting predator-prey relationships, competitive dynamics, and mutualistic interactions.

Phenological mismatches occur when interacting species respond differently to climate cues, disrupting ecological relationships. For example, when insects emerge earlier due to warming but the birds that feed on them don't adjust their migration timing proportionally, both species suffer—insects face less predation control and birds face food shortages. Similar mismatches affect plant-pollinator relationships, predator-prey dynamics, and host-parasite interactions across ecosystems (Kharouba et al., 2018).

Extreme events like wildfires, hurricanes, floods, and droughts exceed many ecosystems' historical disturbance regimes. While many ecosystems evolved with periodic disturbances, changes in frequency, intensity, and combinations of these events can exceed adaptive capacities. For example, forests adapted to occasional fire may not recover when fire return intervals shrink below the time needed for post-fire regeneration, potentially triggering conversion to non-forest ecosystems.

Compounding non-climate stressors like habitat fragmentation, pollution, invasive species, and resource extraction reduce ecosystem resilience to climate impacts. These existing pressures create "adaptation deficits" that make even modest climate changes potentially devastating. For example, coral reefs stressed by pollution, overfishing, and physical damage show dramatically reduced capacity to recover from bleaching events associated with marine heat waves.

Ecosystem adaptation strategies span a spectrum from minimal intervention to intensive management:

Protected area networks preserve habitat, migration corridors, and climate refugia—areas where climate change proceeds more slowly due to topographic, hydrological, or other buffering factors. Expanding and connecting protected areas creates space for species to adapt and migrate as conditions change. This strategy works best when protected areas span elevation gradients, encompass diverse microclimates, and form connected networks rather than isolated islands of habitat.

Ecological restoration repairs degraded ecosystems to enhance their adaptive capacity through improved habitat quality, species diversity, and functional redundancy. Climate-informed restoration increasingly emphasizes process restoration (like natural hydrological regimes) and functional diversity rather than rigid historical composition targets. This approach recognizes that while we can't recreate historical ecosystems under changed climate conditions, we can restore ecological processes and functions that support adaptation.

Assisted migration involves deliberately moving species to areas predicted to provide suitable habitat under future climate conditions but beyond the species' natural dispersal capacity. This controversial approach raises questions about ecological risks, decision authority, success criteria, and resource prioritization. Yet for some climate-threatened species with limited dispersal ability and specific habitat requirements, assisted migration may represent the only alternative to extinction.

Ecological transformation management guides ecosystem transitions when climate change makes maintaining historical systems impossible. Rather than futilely resisting inevitable changes, this approach focuses on maintaining ecological functions, biodiversity, and ecosystem services through transition periods. It might involve facilitating change toward novel but functioning ecosystems rather than attempting to preserve historical communities that climate conditions no longer support.

Ex situ conservation preserves genetic material, seeds, or individuals of climate-threatened species in botanical gardens, seed banks, zoos, and other facilities. While this approach cannot preserve ecosystems or ecological interactions, it maintains options for future restoration efforts and prevents complete extinction of highly vulnerable species. It serves as an insurance policy against worst-case scenarios while other adaptation strategies are pursued.

Ecosystem adaptation raises profound questions about goals and values:

What are we adapting to? Climate projections contain significant uncertainty, especially at the local scales most relevant for ecosystem management. Adaptation strategies must address not just central tendency projections but also extremes, variability, and potential surprises. This uncertainty complicates decision-making about interventions with long-term ecological consequences.

What are we adapting for? Traditional conservation emphasized preserving historical ecological conditions, but climate change increasingly renders this approach unviable. Alternative goals include maximizing biodiversity, maintaining ecosystem services, preserving evolutionary potential, or ensuring ecological function—each leading to different adaptation priorities and interventions.

Who decides? Ecosystem adaptation decisions affect multiple stakeholders with diverse values, knowledge systems, and interests— from indigenous communities with millennia of relationship to particular landscapes to recreational users to extractive industries to future generations. Governance approaches that navigate these diverse perspectives while incorporating both scientific and traditional ecological knowledge remain underdeveloped in many regions.

Despite these challenges, ecosystem adaptation offers unique advantages and co-benefits compared to other adaptation domains. Healthy ecosystems provide natural infrastructure that buffers climate impacts— wetlands that reduce flooding, forests that moderate temperature extremes, coastal ecosystems that protect against storms and sea level rise. These "nature-based solutions" often prove more cost-effective and resilient than engineered alternatives while providing multiple co-benefits from carbon sequestration to habitat provision to recreational opportunities.

Ultimately, ecosystem adaptation requires humility, recognizing that we neither fully understand nor control the complex systems we seek to help navigate rapid change. This doesn't mean abandoning management responsibility but rather adopting approaches that work with rather than against ecological processes, preserve options for the future, and balance intervention with allowing nature's own adaptive capacity to operate where possible.

Managed Retreat: When Adaptation Has Limits

While adaptation offers many pathways for reducing climate vulnerability, we must confront an uncomfortable reality: some impacts exceed adaptation limits. In certain locations and scenarios, no amount of seawalls, air conditioning, irrigation, or other adaptations can maintain viable human settlements or productive ecosystems. When this occurs, managed retreat—the planned relocation of people, assets, and activities from areas of intolerable risk—becomes necessary (Siders et al., 2019).

Managed retreat becomes necessary in several scenarios:

Sea level rise will eventually inundate many coastal areas regardless of protection efforts. While seawalls and other barriers can delay inundation, they eventually fail, require prohibitive maintenance, or create other problems like groundwater salinization and ecosystem destruction. For low-lying islands, coastal plains, and river deltas facing multiple feet of sea level rise this century and beyond, managed retreat represents the only long-term sustainable option for at least portions of these regions (Oppenheimer et al., 2019).

Increasing flood frequency from more intense precipitation renders some riverside and floodplain areas increasingly uninhabitable. When "100-year floods" become annual or biennial events, the economic and social viability of affected areas deteriorates regardless of flood protection infrastructure. Many communities already face this reality, with repetitive flood losses exceeding property values and making continued occupation untenable.

Escalating wildfire risk transforms formerly manageable fire regimes into existential threats in the wildland-urban interface. As climate change extends fire seasons, increases fire intensity, and creates new fire weather patterns, some development patterns become fundamentally unsafe regardless of building codes or fuels management. These changing conditions force reconsideration of settlement patterns and land uses in fire-prone landscapes.

Water insecurity renders some agricultural regions and settlements non-viable as groundwater depletion, reduced snowpack, and changing precipitation patterns permanently alter water availability. When irrigation becomes impossible or municipal water sources fail, managed retreat from affected areas becomes inevitable despite any efficiency improvements or conservation efforts. This reality already faces communities from California's Central Valley to the Middle East to Australia's Murray-Darling Basin.

Extreme heat approaches or exceeds human physiological limits in some regions, particularly when combined with humidity that prevents evaporative cooling (sweating). When wet-bulb temperatures regularly exceed thresholds where even healthy humans cannot survive outdoors, and where energy-intensive cooling remains unavailable or unreliable, previously habitable areas become effectively unlivable. This threshold approaches more rapidly than many climate projections initially suggested.

Despite its necessity in these scenarios, managed retreat faces substantial challenges:

Financial barriers arise from property values tied to locations increasingly recognized as non-viable. Homeowners resist abandoning their largest financial assets without adequate compensation, while governments struggle to fund large-scale buyout programs. Insurance markets increasingly recognize these risks through coverage limitations and premium increases, but financial mechanisms for equitable retreat remain underdeveloped in most jurisdictions.

Cultural attachments to place transcend economic considerations, particularly for communities with deep historical connections to specific locations. Indigenous communities, multi-generational residents, and culturally distinct groups often form identities inextricably linked to particular geographies. For these communities, managed retreat represents not just physical relocation but existential threats to cultural continuity and collective identity.

Receiving area constraints limit relocation options as climate impacts affect multiple regions simultaneously. When considering managed retreat from coastal flooding, for example, obvious inland alternatives often face their own climate challenges from extreme heat to water limitations to wildfire risk. This spatial compression of viable settlement areas creates both economic pressures (higher land costs in safer areas) and social tensions as different displaced populations compete for limited relocation options.

Governance gaps emerge as existing institutions prove ill-equipped for managing large-scale, long-term relocation processes. Traditional disaster management focuses on recovery and rebuilding in place rather than strategic retreat and resettlement. Meanwhile, property rights frameworks, land use regulations, infrastructure financing mechanisms, and intergovernmental coordination processes all present barriers to coherent retreat planning and implementation.

Despite these challenges, several approaches show promise for making managed retreat more viable:

Phased implementation allows gradual transition rather than abrupt displacement. This might involve prohibiting new development or major repairs after damage while allowing continued occupancy during a multi-year or multi-decade transition period. This approach provides time for community planning, infrastructure development in receiving areas, and individual adjustment while still maintaining a clear trajectory toward eventual retreat from untenable locations.

Buyout programs with community co-design address both economic and social dimensions of retreat. The most successful programs combine fair compensation with community involvement in decisions about timing, process, and destination options. When affected communities participate meaningfully in program design and implementation, both economic outcomes and social cohesion improve compared to top-down approaches.

Receiving area preparation integrates retreating populations into new locations through coordinated housing, infrastructure, economic development, and social inclusion initiatives. Rather than simply relocating households individualistically, comprehensive approaches consider community integrity, livelihood opportunities, service access, and cultural continuity in receiving areas. This perspective views retreat as community adaptation rather than mere displacement.

Legal innovations create new frameworks for managing retreat through mechanisms like rolling easements that gradually shift property rights as conditions change, transferable development rights that compensate property owners for relocation, and new jurisdictional arrangements for communities that move cohesively rather than dispersing. These approaches create legal pathways for orderly retreat that protect both individual rights and community interests.

While managed retreat represents adaptation's most challenging frontier, it also offers opportunities for rethinking fundamental aspects of human settlement patterns, property regimes, community structures, and relationships with place. Rather than merely reacting to loss, forward-thinking retreat initiatives reimagine how communities might reorganize in more climate-resilient configurations—potentially creating settlements better adapted to future conditions than those being left behind.

Retreat's necessity in some contexts doesn't imply surrender in the broader climate struggle. Rather, strategic retreat from truly untenable locations allows concentration of adaptation resources where they can effectively reduce vulnerability elsewhere, while avoiding endless cycles of disaster, rebuilding, and repeat disaster in locations facing insurmountable climate risks. It's the climate equivalent of the military maxim that sometimes you must retreat from one position to defend another more effectively—a tactical adjustment serving strategic goals rather than outright defeat.

The Mitigation-Adaptation Balance: Both/And, Not Either/Or

Throughout this chapter, we've focused on adaptation strategies while occasionally noting their connection to mitigation (emissions reduction). But before concluding, we need to directly address this relationship between adaptation and mitigation—two essential, complementary responses to climate change that work best in careful balance rather than isolation or opposition (IPCC, 2022).

Several principles guide this balanced approach:

Both are necessary, neither is sufficient: No realistic level of mitigation can prevent all climate impacts given changes already underway from historical emissions. Similarly, no level of adaptation can address unbounded climate change under continued high emissions. The either/or framing of adaptation versus mitigation represents a fundamental misunderstanding of climate dynamics and response options. The real question isn't which to pursue but how to optimize their combination.

Timing considerations differ: Mitigation benefits manifest gradually over decades through avoided climate change, while adaptation benefits often appear more immediately through reduced vulnerability to current and near-term impacts. This temporal difference creates political challenges, as adaptation may seem more urgent and tangible while mitigation's benefits remain more distant and diffuse. Balanced approaches recognize both immediate adaptation needs and longer-term mitigation imperatives rather than emphasizing either timeframe exclusively.

Geographic scales vary: Mitigation offers global benefits regardless of emission location, while adaptation primarily benefits the specific locations where measures are implemented. This spatial difference affects incentive structures, as mitigation represents a global public good subject to free-rider problems, while adaptation benefits are more directly captured by those investing in it. Addressing these different incentive structures requires coordinated international frameworks for mitigation alongside context-specific adaptation approaches.

Sectoral emphases differ: Some sectors present primarily mitigation challenges (like energy systems), others primarily adaptation challenges (like coastal management), and many require integrated approaches addressing both dimensions simultaneously (like agriculture and forestry). Balanced climate responses recognize these sectoral differences rather than applying one-size-fits-all approaches across the economy.

Several approaches help optimize the mitigation-adaptation balance:

Low-regrets options that simultaneously advance both adaptation and mitigation goals. Examples include regenerative agriculture that both sequesters carbon and improves drought resilience; ecosystem restoration that provides both carbon sinks and natural infrastructure for climate adaptation; and green buildings that reduce both energy consumption and vulnerability to temperature extremes. These win-win approaches deserve particular emphasis in climate planning and investment.

Avoid adaptation-mitigation tradeoffs where possible by identifying and addressing conflicts before they become locked in. Examples include ensuring that adaptation infrastructure (like seawalls or water treatment facilities) minimizes energy consumption and emissions; avoiding adaptation paths that increase land use emissions (like deforestation for agricultural expansion under changing climate conditions); and designing energy transitions with resilience to climate impacts built in from the start.

Risk management approaches that explicitly consider both mitigation and adaptation within integrated climate responses. Rather than treating these as separate domains handled by different institutions, risk management frameworks assess how combinations of mitigation and adaptation measures address overall climate risk profiles. This integrated perspective supports more strategic resource allocation across both response categories.

Transformational thinking that reimagines systems to simultaneously achieve dramatic emissions reductions and fundamental resilience improvements. Rather than incremental tweaks to existing systems that address adaptation and mitigation separately, transformational approaches seek new configurations that achieve both goals through more fundamental redesign. Examples include regenerative circular economies, climate-smart urban systems, and carbon-negative, climate-resilient food systems.

This balanced perspective doesn't imply equal resource allocation between adaptation and mitigation in all contexts. The appropriate balance varies with geographic context (some locations face more immediate climate threats than others), development stage (adaptation often takes higher priority in vulnerable developing countries), and sector (energy systems emphasize mitigation while coastal management emphasizes adaptation). The key principle isn't rigid formula but thoughtful integration based on specific contexts and needs.

What this balanced approach does require is overcoming the false narrative that adaptation represents giving up on mitigation or accepting climate change as inevitable and unmodifiable. Adaptation doesn't signal surrender in the climate struggle—it represents realistic engagement with both present vulnerabilities and future risks. The strongest climate leadership embraces both adaptation and mitigation as essential, complementary responses working in concert rather than competition.

As we navigate increasingly turbulent climate conditions in coming decades, this both/and perspective will prove increasingly vital. The communities, countries, and companies that thrive will be those that simultaneously reduce emissions to limit future climate change while building resilience to impacts already underway or locked in by past emissions. Neither response alone suffices, but together they offer our best pathway through the climate challenges ahead.

Conclusion: Adapting Forward, Not Giving Up

We've covered substantial ground in this exploration of climate adaptation—from infrastructure and agriculture to public health and ecosystems, from managed retreat and adaptation finance to justice considerations and inherent limits. Throughout this journey, one theme consistently emerges: adaptation represents not resignation but resilience, not surrender but strategic adjustment to changing conditions while maintaining core values and goals (National Research Council, 2010).

This "adapting without giving up" mindset offers several key insights:

Adaptation is active, not passive: Rather than merely accepting climate impacts as they occur, adaptation actively reshapes systems, structures, and practices to reduce vulnerability and enhance resilience. It represents agency in the face of changing conditions—the capacity to influence outcomes even when we cannot fully control the driving forces. This active orientation challenges fatalistic narratives that portray adaptation as merely reactive submission to inevitable change.

Adaptation preserves what matters most: While adaptation often involves significant changes to how we do things, it aims to preserve what matters most through these transitions—human wellbeing, ecological function, cultural continuity, and core values. Like a jazz musician improvising on a theme, adaptation maintains essential elements while creatively responding to changing conditions. This preservation-through-change perspective offers alternatives to both rigid resistance and complete surrender.

Adaptation creates opportunity alongside challenge: While climate impacts create undeniable hardships, adaptation processes can catalyze beneficial changes that might otherwise face resistance—from more sustainable resource management to more inclusive governance to more integrated planning across sectors and jurisdictions. These co-benefits don't negate climate damage but do reveal possibilities for positive transformation within necessary responses to changing conditions (Adger et al., 2017).

Adaptation complements rather than replaces mitigation: As we've emphasized throughout, adaptation works alongside emissions reduction rather than substituting for it. The most effective adaptation occurs within climate scenarios moderated by ambitious mitigation, while the most effective mitigation includes adaptation to changes already underway or locked in by past emissions. This complementary relationship means adaptation strengthens rather than weakens the case for transformative climate action.

Looking ahead, several adaptation frontiers deserve particular attention:

Governance innovation for conditions beyond historical experience, where traditional decision systems built around stationarity and historical analogs no longer suffice. New governance approaches incorporating deep uncertainty, multiple knowledge systems, collaborative learning, and adaptive management show promise for navigating unprecedented conditions without either paralysis or overconfidence.

Social-ecological resilience that integrates human and natural systems rather than treating them as separate domains. This perspective recognizes that human wellbeing depends on functional ecosystems, while ecological resilience increasingly depends on human decisions and management. Their futures are inextricably linked, requiring adaptation approaches that address these interdependencies rather than optimizing either system in isolation.

Adaptation justice frameworks that center equity considerations in adaptation planning, financing, and implementation. As climate impacts intensify, ensuring that adaptation benefits reach those facing greatest vulnerability—and that adaptation costs don't fall disproportionately on those least responsible for climate change—becomes both a moral imperative and a practical necessity for effective climate resilience.

Transformational adaptation beyond incremental adjustments to existing systems when climate changes exceed the adaptive capacity of current configurations. These deeper transformations—whether in settlement patterns, agricultural systems, economic structures, or governance arrangements—require both technical capacity and social processes that support navigating disruptive change while maintaining social cohesion and core values.

Perhaps most importantly, adaptation requires hope—not naive optimism that ignores difficult realities, but clear-eyed hope grounded in both honest assessment of challenges and genuine commitment to addressing them. Without hope, adaptation efforts falter in the face of overwhelming projections. With it, even daunting climate scenarios become navigable through determined, creative collective action.

This hope-fueled adaptation doesn't mean everyone and everything will emerge unscathed from climate change. Some losses appear unavoidable despite our best efforts. But between the fantasy that nothing will change and the fatalism that everything valuable will be lost lies the realistic middle ground where adaptation operates—acknowledging unavoidable changes while actively shaping which values, systems, and relationships persist through transformation.

In this sense, adapting without giving up means maintaining commitment to core human and ecological values even as the means of sustaining them evolve. It means facing climate realities clear-eyed while refusing surrender to despair. And it means recognizing that while we cannot preserve everything exactly as it has been, we retain tremendous agency in determining what endures, what transforms, and what emerges anew as we navigate this unprecedented planetary transition together.

In the next chapter, we'll explore how younger generations are leading climate action with distinctive perspectives, approaches, and moral clarity that offer lessons for all engaged in this defining challenge of our time.

Chapter 14: The Youth Know More Than You Think

When the Kids Take Charge

Remember when your parents used to say, "You'll understand when you're older"? Well, when it comes to climate change, we're witnessing a fascinating role reversal. Today's youth aren't waiting to grow up to understand the climate crisis—they're already leading the charge to address it. And rather than waiting for adults to catch up, they're pushing forward with remarkable clarity, urgency, and vision that many of their elders seem to lack.

The rise of youth climate activism represents one of the most significant social developments in the climate sphere. From school strikes that spread globally to youth-led litigation challenging government inaction, from social media campaigns that mobilize millions to community projects creating local solutions, young people are reshaping the climate conversation and forcing action at all levels (Fisher and Nasrin, 2021).

This youth leadership isn't some cute sideshow or performative activism —it's proving to be one of the most effective forces for climate progress. Young activists have shifted public discourse, influenced policy, pressured institutions to divest from fossil fuels, and inspired broader generational engagement with climate issues. They've accomplished this despite limited traditional power, resources, or political access, demonstrating that when it comes to moral clarity and mobilization capacity, youth bring unique strengths to climate action (Rapid Transition Alliance, 2021).

In this chapter, we'll explore the youth climate movement—its distinctive characteristics, approaches, accomplishments, and the lessons it offers for climate engagement across generations. We'll examine how young people are simultaneously inheriting a climate crisis not of their making and actively shaping how society responds to it. And we'll consider why, when young people speak about climate change, the rest of us would be wise to listen.

The Rise of Youth Climate Activism: Not Just Skipping School

The contemporary youth climate movement emerged from a perfect storm of factors: accelerating climate impacts, mounting scientific urgency, political inaction, and new communication technologies that enable unprecedented connectivity. While youth environmental activism has existed for decades, the current movement's scale, visibility, and impact represent something genuinely new.

Several key moments mark this movement's evolution:

Fridays for Future began in August 2018 when 15-year-old Greta Thunberg sat alone outside the Swedish parliament with a handmade sign reading "School Strike for Climate." What started as one teenager's solitary protest expanded into a global movement involving millions of students across more than 150 countries. The simple tactic—skipping school on Fridays to demand climate action—created a powerful symbolism: why pursue education for a future that climate change threatens to undermine? It's like studying for a test that's been canceled —a rational reallocation of priorities in the face of existential threat (Fridays For Future, 2023).

Zero Hour launched in 2017 when then-15-year-old Jamie Margolin founded a youth-led organization focused on centering environmental justice within climate activism. The group organized the "Youth Climate March" in Washington, D.C. and other cities in July 2018, highlighting how climate impacts disproportionately affect marginalized communities. This justice-centered approach recognized from the beginning what many mainstream environmental movements had sometimes neglected—that climate change isn't just an environmental issue but a profound social justice challenge (Zero Hour, 2024).

Sunrise Movement emerged in 2017 as a youth-led organization combining direct action, electoral engagement, and policy advocacy around climate change. Their occupation of House Speaker Nancy Pelosi's office in November 2018—joined by newly-elected Representative Alexandria Ocasio-Cortez—helped catapult the Green New Deal into national consciousness. This strategic combination of disruptive protest, media-savvy messaging, and practical policy proposals demonstrated sophisticated political understanding belying the movement's young age (Sunrise Movement, 2023).

Youth climate litigation represents another powerful approach, with young plaintiffs suing governments for failing to protect their constitutional rights to life, liberty, and a stable climate. Cases like Juliana v. United States and similar suits worldwide use legal systems to establish governmental responsibility for climate action. These cases effectively translate moral arguments about intergenerational justice into legal claims, creating new pathways for accountability when traditional political channels prove unresponsive (Our Children's Trust, 2025).

Local youth initiatives complement these high-profile movements with community-level action addressing specific climate impacts or solutions. From urban gardening projects in food deserts to renewable energy advocacy in schools to plastic pollution awareness campaigns, these grassroots efforts demonstrate how global concerns translate into local action. Such projects often receive less media attention than international movements but create tangible impacts while developing young leaders' skills and agency.

What unites these diverse approaches is their youth leadership, moral framing, impatience with incremental change, and skillful navigation of both traditional and new media environments. Unlike most adult-led climate initiatives, youth activism typically emphasizes the fundamental injustice of burdening future generations with climate impacts created by past and current generations. This intergenerational frame transforms climate change from primarily a technical or economic challenge into a profound moral issue about obligations to young people and future generations (The Conversation, 2020).

Contrary to some dismissive characterizations, these movements don't reflect naive idealism disconnected from practical realities. The most effective youth climate initiatives combine moral clarity with sophisticated strategic thinking, media savvy, policy understanding, and organizational skill. Their leaders speak articulately about climate science, energy systems, political constraints, economic transitions, and social justice dimensions—often with greater command of these interdisciplinary connections than many adult "experts" siloed within narrow specialties.

What these youth lack in traditional political power, they compensate for with creativity, authenticity, digital fluency, and ability to mobilize peer networks. Their movements operate both within and outside conventional channels—voting and lobbying elected officials where possible, while simultaneously building independent pressure through protests, strikes, boycotts, and media campaigns. This multi-tactical approach provides flexibility unavailable to many established institutions constrained by tradition, precedent, or institutional interests.

Why Now? The Perfect Storm for Youth Leadership

The emergence of powerful youth climate movements at this particular historical moment isn't coincidental. Several converging factors created fertile ground for youth leadership to flourish in ways unimaginable even a decade earlier.

Climate reality becoming undeniable as impacts intensify globally. Today's youth are the first generation witnessing clear climate disruption during their formative years rather than as abstract future projections. For many young people, climate change isn't a theoretical concern but a lived reality through more intense storms, heat waves, floods, and fires affecting their communities. This experiential relationship with climate change creates urgency that theoretical understanding alone rarely generates.

Science communication improvements have made complex climate information more accessible to non-specialists, including young people. Better visualizations, clearer explanations, and more effective translation of technical concepts into everyday language have enabled younger generations to understand climate science despite limited formal scientific training. Organizations like NASA and NOAA have created youth-oriented climate resources, while social media platforms spread accessible climate information through formats resonating with younger audiences.

Digital connectivity enables unprecedented youth organization and amplification. Previous generations facing global challenges lacked tools to coordinate globally, share strategies, or amplify messages without significant institutional support. Today's digital platforms allow youth movements to form, coordinate, and grow with minimal resources beyond smartphones and internet connections. A compelling message or innovative tactic can spread worldwide within days, creating rapid knowledge sharing and movement building that previous generations couldn't achieve (Fridays For Future, 2023).

Intergenerational climate injustice becomes increasingly stark as climate projections clarify impacts across timescales. More sophisticated climate models and impact assessments reveal how today's policy decisions create consequences extending centuries into the future. This temporal perspective highlights the profound unfairness of climate change—where those contributing least to the problem (young people and future generations) face its worst consequences. This injustice provides powerful moral motivation for youth engagement.

Political failures around climate action despite decades of scientific warnings and international negotiations. Many young climate activists were born after the 1992 Rio Earth Summit established the UN Framework Convention on Climate Change. They've witnessed their entire lives within a context of scientific certainty about climate threats alongside political inability to address these threats adequately. This combination of clear problem identification with ineffective response creates perfectly justified impatience with conventional political processes.

Educational improvements around environmental issues provide stronger foundations for youth engagement. Many school systems have incorporated climate science, sustainability principles, and systems thinking into curricula, creating better-informed young people compared to previous generations. This educational foundation enables more sophisticated engagement with climate issues, building on formal learning rather than requiring youth to educate themselves entirely outside educational institutions.

Cultural shifts in parenting and education increasingly encourage youth voice and agency rather than strict hierarchical relationships. While significant geographic and cultural variations exist, many societies show greater openness to youth perspectives and leadership compared to previous generations. These shifts create more space for authentic youth engagement on social issues rather than rigid expectations that young people should remain silent until adulthood.

These factors combine to create unprecedented conditions for youth climate leadership—providing both motivation (clear climate threats to their future) and means (digital connectivity, scientific accessibility, cultural openness) for effective engagement. The resulting movements reflect both these enabling conditions and young people's distinctive perspectives, priorities, and approaches to climate challenges.

Importantly, these movements don't merely replicate adult environmental activism with younger participants. They bring genuinely different frameworks—emphasizing intergenerational justice, systems change, intersectionality, and urgency in ways that many established environmental organizations historically neglected. They're not just younger versions of existing approaches but represent qualitatively different engagement with climate challenges—often more holistic, justice-centered, and transformational than conventional environmental politics (Resilience, 2021).

The Youth Climate Perspective: What Makes It Different

Youth climate activism doesn't just differ from adult-led movements in terms of participants' ages—it brings distinctive frameworks, priorities, and approaches that reflect young people's unique position in relation to climate change. Understanding these differences helps explain both the movements' rapid growth and their catalytic effect on broader climate politics.

Several elements characterize the distinctive youth climate perspective:

Future-oriented timeframes shape youth climate engagement differently than adult perspectives. For today's teenagers and young adults, 2050 climate projections don't represent some distant hypothetical—they describe midlife conditions they'll personally experience. Similarly, 2100 scenarios concern their children's and grandchildren's lives rather than abstract future generations. This collapsed psychological distance creates different risk perceptions and urgency compared to older adults whose decision-making often prioritizes shorter timeframes.

Moral clarity characterizes youth climate framing, which typically presents climate inaction as a clear ethical failure rather than a complex balancing of competing priorities. This perspective doesn't reflect naivety but rather recognition that core values like intergenerational responsibility, justice, and preservation of life-sustaining ecological conditions should take precedence over short-term economic considerations. When young activists declare "our house is on fire," they're expressing moral clarity about prioritization rather than denying complexity (Amnesty International, 2019).

Systems thinking appears more naturally in youth climate frameworks compared to many adult approaches. Rather than addressing climate change as an isolated environmental issue, youth movements typically emphasize its connections with economic systems, social justice, racial equity, colonialism, and other structural factors. This integrated perspective rejects siloed approaches in favor of transformational change addressing root causes rather than symptoms. Young activists consistently emphasize that climate change reflects systemic failures requiring systemic solutions, not merely technological fixes within existing frameworks (MDPI, 2020).

Justice-centered framings place equity at the core rather than periphery of climate action. Youth movements typically emphasize that effective climate solutions must address historical and continuing injustices that create disproportionate vulnerability. This approach recognizes that both climate impacts and transition challenges affect communities unequally based on race, class, geography, and other factors. Rather than treating these justice dimensions as secondary considerations or potential co-benefits, youth climate frameworks identify them as central to legitimate and effective climate action (PubMed Central, 2021).

Psychological openness to transformational change distinguishes youth perspectives from older generations more invested in existing systems. Having spent less time within current economic, social, and political arrangements, young people typically show greater willingness to reimagine these systems rather than accepting their fundamental structure as given. This flexibility creates more space for transformational thinking unbounded by assumptions about what's politically feasible within existing constraints.

Digital nativity shapes both organizing tactics and communication strategies in youth movements. Having grown up with social media, smartphones, and distributed digital networks, young activists navigate these environments with intuitive fluency that many older organizations struggle to match. This digital fluency enables rapid information sharing, distributed organizing without centralized control, creative media strategies, and global coordination that significantly amplifies movement impacts beyond what their limited resources might otherwise achieve.

Scientific literacy combines with moral framing in contemporary youth climate movements. Unlike some previous youth-led movements, today's climate activism demonstrates sophisticated understanding of scientific evidence alongside moral arguments. Young activists routinely cite specific IPCC reports, emissions pathways, carbon budgets, and climate impacts with remarkable precision, grounding moral claims in scientific understanding rather than treating them as separate domains (Global Center on Adaptation, 2024).

Institutional skepticism reflects young people's lived experience with institutional climate failure. Today's youth have witnessed their entire lives within a context of scientific clarity about climate threats alongside institutional inability to address these threats effectively. This history naturally produces skepticism toward claims that the same institutions and processes that created climate inaction for decades will suddenly deliver adequate solutions without significant pressure and transformation.

These distinctive elements combine to create youth climate engagement that differs qualitatively from most adult-led climate initiatives. Rather than representing merely enthusiasm without experience, these characteristics reflect genuine insights arising from young people's unique historical position. They've grown up simultaneously more aware of climate threats and less constrained by assumptions about what's possible than previous generations, creating space for fresh perspectives on challenges that have often seemed intractable within conventional frameworks.

The youth climate perspective doesn't claim superior wisdom about all aspects of climate solutions—it acknowledges the value of technical expertise, institutional knowledge, and practical experience that older generations contribute. But it does insist that moral clarity, systems thinking, justice centering, psychological openness, and digital fluency represent genuine strengths rather than deficiencies to be outgrown with maturity. These elements don't reflect naive idealism but rather clear vision about both problems and potential solutions that existing approaches have often failed to recognize or prioritize.

What They've Accomplished: More Than Just Noise

Critics sometimes dismiss youth climate activists as creating more noise than impact—staging theatrical protests while lacking power to implement actual solutions. This characterization fundamentally misunderstands both movement strategy and the significant accomplishments youth climate leadership has already achieved across multiple domains.

The youth climate movement has delivered substantial results in several areas:

Discourse transformation represents perhaps the most significant youth climate accomplishment. Young activists have fundamentally altered public conversation about climate change—shifting emphasis from technical debates about emissions percentages and target dates toward moral questions about intergenerational responsibility and systemic change. This discourse shift appears in media coverage, political platforms, corporate communications, and everyday conversations, reflecting how youth framing has permeated broader climate discussions. It's like changing the questions on the test—once the framing shifts from "What's economically feasible?" to "What's morally necessary for our children's future?", the answers change too (Sustainability Science, 2023).

Institutional commitments accelerated dramatically following youth climate mobilizations. Universities, pension funds, religious organizations, foundations, and governments announced divestment from fossil fuels, net-zero targets, climate emergency declarations, and policy changes following sustained youth pressure. These institutional responses often explicitly acknowledged youth movements as catalysts, with decision-makers citing youth protests, open letters, or direct engagement as factors motivating their actions. While implementation of these commitments remains uneven, they create accountability mechanisms and political pressure previously absent.

Electoral influence expanded as youth climate movements translated moral energy into political power. In multiple countries, youth climate activism corresponded with increased youth voter turnout and climate prioritization in electoral politics. Candidates explicitly addressing climate concerns performed better among younger voters, creating new political incentives for climate leadership. This electoral translation doesn't represent abandonment of movement tactics but rather strategic supplementation of protest with voting power—recognizing that both pressure and participation remain necessary for systemic change.

Policy frameworks like the Green New Deal gained prominence and political traction through youth movement support. These comprehensive approaches linking climate action with economic transformation and social justice might have remained marginal without youth movements amplifying them through protests, media engagement, and electoral pressure. By providing organized constituency support for ambitious policy frameworks, youth climate activists helped shift political calculations about what proposals deserve serious consideration rather than dismissal as unrealistic (Sunrise Movement, 2023).

Community implementations demonstrate youth leadership beyond protest and advocacy. Young people have initiated renewable energy projects, urban gardens, circular economy initiatives, climate education programs, and other concrete implementations addressing both climate mitigation and adaptation. These projects complement broader movement-building by creating tangible examples of solutions while developing practical skills, community relationships, and implementation experience. They represent climate leadership beyond criticism of existing systems to active creation of alternatives.

Legal precedents established through youth-led climate litigation have expanded recognition of rights to a stable climate and government responsibilities to protect these rights. While final legal outcomes remain pending in many cases, these proceedings have already created valuable judicial opinions, public education, and political pressure that complement other movement strategies. By translating moral arguments into legal claims, these cases create new accountability pathways when traditional political channels prove unresponsive (Grist, 2025).

Cultural shifts around climate engagement reflect youth influence beyond formal politics. The normalization of climate concern, personal climate action, and climate-conscious decision-making among broader populations shows imprints of youth climate framing. From sustainable consumption choices to career priorities to financial decisions, climate considerations have entered mainstream culture partly through youth-led conversations that expanded beyond activist circles to influence broader social norms.

These diverse accomplishments demonstrate sophisticated understanding of social change pathways beyond simplistic models where only direct policy control constitutes real impact. Youth climate movements recognize that discourse transformation, institutional commitments, electoral influence, community implementation, legal precedents, and cultural shifts create multiple pressure points within complex systems. Rather than fixating solely on legislative victories— especially when facing political systems often unresponsive to their concerns—youth climate leadership works simultaneously across these multiple domains to create cumulative impact exceeding what any single approach could achieve.

The tangible results already visible contradict dismissive characterizations of youth climate activism as merely symbolic or performative. These movements have demonstrably altered institutional behavior, political calculations, legal frameworks, and cultural norms within timeframes that many established environmental organizations proved unable to match despite greater resources and institutional access. This track record suggests not naive idealism but rather strategic effectiveness in catalyzing change through multiple complementary pathways rather than relying on singular approaches easily blocked within existing power structures.

Criticisms and Challenges: Growing Pains or Fundamental Flaws?

Like any social movement, youth climate activism faces criticisms and challenges. Some represent growing pains typical of rapidly expanding movements, while others reflect more fundamental tensions requiring thoughtful navigation. Understanding these criticisms—both legitimate concerns and mischaracterizations—helps assess the movement's evolution and potential future trajectories.

Privilege critiques highlight how visible youth climate leadership often comes from relatively privileged backgrounds despite climate impacts disproportionately affecting marginalized communities. The most internationally recognized youth activists typically have educational, economic, and social advantages enabling their activism through resources, flexible schedules, media access, and safety from repression. This pattern raises valid questions about representation and whose perspectives shape movement priorities. It's like only interviewing honors students about educational reform while ignoring those struggling within the system who might offer different insights (Taylor & Francis Online, 2020).

Pragmatism tensions emerge between moral absolutism and practical implementation requirements. Some youth climate rhetoric employs categorical moral framing that can complicate negotiation, compromise, and incremental progress necessary within complex political systems. When "climate justice now" serves as both rallying cry and policy demand, translating between movement energy and implementable steps sometimes proves challenging. It's like demanding an A+ when the current grade is D—righteous but potentially leaving insufficient space for acknowledging meaningful progress from D- to C or B.

Technical capacity questions arise regarding specialized knowledge needed for effective climate solutions. While youth movements demonstrate impressive scientific literacy about climate problems, designing and implementing solutions often requires technical expertise in energy systems, policy design, finance, and other domains where age-related experience gaps naturally exist. Balancing fresh perspective with necessary technical knowledge represents an ongoing challenge. It's like having brilliant diagnosis of a medical condition without equivalent expertise in treatment protocols.

Sustainability challenges affect movements requiring intense commitment amid limited resources. The emotional demands of climate activism create burnout risks, especially for young people simultaneously managing educational requirements, economic precarity, and normal developmental tasks. Movements built on exceptional commitment from extraordinary individuals face challenges maintaining momentum as personal circumstances change. It's like sprinting the first mile of a marathon—inspiring but potentially unsustainable for the full distance required.

Tactical disagreements emerge within youth movements regarding appropriate approaches. While media coverage often presents youth climate activism as monolithic, significant internal debates exist about protest tactics, electoral engagement, institutional collaboration, and theory of change. These disagreements reflect normal movement dynamics rather than dysfunction but sometimes create confusion about movement priorities and potential partnership opportunities. It's like siblings who generally agree about family problems but differ dramatically on how to address them.

Transition challenges arise as youth movements institutionalize and leaders age. Movements beginning with spontaneous energy face questions about organizational structure, decision-making processes, leadership transitions, and maintaining youth-centered perspectives as initial leaders themselves age into adulthood. These lifecycle questions affect all movements but particularly those explicitly organized around generational identity. It's like the awkward transition when a childhood club tries to create formal bylaws and leadership roles after years of informal operation.

While these challenges require thoughtful navigation, several criticisms of youth climate activism reflect misunderstanding or mischaracterization rather than legitimate concerns:

"Just children" dismissals attempt to delegitimize youth perspectives based solely on age rather than substantive engagement with their arguments. When critics focus on youth activists' age rather than their actual claims, they commit an ad hominem fallacy that reveals argumentative weakness rather than meaningful critique. The question isn't whether activists are young but whether their analysis is accurate and their proposals warranted. It's like dismissing Einstein's physics papers because he worked in a patent office rather than addressing his actual equations.

"Manipulated pawns" accusations suggest young activists lack autonomous agency and merely parrot adult activism without independent thinking. These characterizations fundamentally misunderstand both contemporary youth development and movement operations. Today's youth climate leaders demonstrate sophisticated independent analysis, strategic thinking, and self-directed organizing belying narratives about adult manipulation. It's like claiming students only learn math because teachers make them—ignoring their intrinsic motivation and independent intellectual engagement (Time, 2022).

"Hypocritical consumers" attacks highlight perceived contradictions between climate advocacy and personal consumption choices within fossil-dependent systems. These critiques misunderstand both movement focus on systemic change rather than individual virtue and the reality that youth have minimal control over energy systems shaping their consumption options. It's like criticizing someone for getting wet while advocating for roof repairs during a rainstorm—the advocacy addresses the cause while the wetness reflects current conditions beyond individual control.

"Climate anxiety" pathologizing attempts to reframe youth climate concern as psychological disorder rather than rational response to genuine threats. While climate distress certainly affects many young people, framing their concern as primarily psychological rather than responding to the substantial threats they face misdiagnoses the situation. The appropriate response to justified concern isn't treating the concerned individuals but addressing the underlying threats creating their concern. It's like treating someone for "fire anxiety" while their house burns rather than extinguishing the flames.

These criticisms often reveal more about critics' discomfort with youth empowerment and substantive climate action than about actual movement weaknesses. The most legitimate challenges facing youth climate activism aren't fundamentally different from those facing any social movement seeking transformational change within complex systems—questions of inclusion, strategy, technical implementation, sustainability, tactics, and institutional evolution that require ongoing reflection and adaptation rather than perfect resolution.

Intergenerational Climate Action: Beyond the Generation Gap

While youth leadership has proven catalytic for climate progress, the most effective climate responses ultimately involve intergenerational collaboration rather than generational division. Different age cohorts bring complementary strengths, knowledge, and capacities to climate challenges that prove more powerful in combination than isolation. Finding pathways for effective partnership while maintaining youth leadership represents a crucial climate movement evolution.

Several approaches show promise for effective intergenerational climate collaboration:

Complementary roles leverage different generational strengths rather than expecting identical engagement across age groups. Youth often excel at mobilization, moral framing, digital organizing, and public communication. Older adults typically bring technical expertise, institutional knowledge, financial resources, and established authority within existing systems. Rather than expecting everyone to contribute identically, effective collaboration acknowledges these different capacities and creates space for each to maximize their distinctive contributions. It's like recognizing that a successful basketball team needs both agile guards and powerful centers rather than expecting everyone to play the same position.

Clear leadership principles establish appropriate decision-making frameworks that respect youth agency while incorporating diverse perspectives. Rather than defaulting to traditional hierarchies where age automatically confers greater authority, effective intergenerational collaboration develops explicit governance that allocates voice and decision rights based on impact, expertise, and representation considerations. These frameworks might look different across contexts but share commitment to meaningful youth leadership rather than mere tokenistic inclusion. It's like family meetings where everyone gets a real vote rather than parents merely pretending to consider children's input before making predetermined decisions.

Knowledge exchange creates bidirectional learning rather than unidirectional teaching from old to young. While older generations certainly possess valuable knowledge from education and experience, youth bring distinctive insights from their generational perspective, digital fluency, and fresh engagement with complex problems. Effective collaboration creates structured opportunities for this bidirectional knowledge sharing rather than assuming knowledge flows only from experienced to inexperienced participants. It's like recognizing that digital natives can teach digital immigrants about technology navigation while benefiting from immigrants' deeper understanding of the pre-digital context that shaped current systems.

Resource sharing addresses power imbalances stemming from differential access to funding, institutional platforms, media relationships, and other resources. Older individuals and organizations often control substantially greater resources than youth-led initiatives despite the latter's innovative approaches and mobilization capacity. Intentional resource sharing—from fiscal sponsorship to office space to media connections—can significantly amplify youth leadership impact without imposing external control through traditional funding dynamics. It's like parents providing allowance without dictating exactly how it must be spent—enabling agency while sharing resources controlled through no particular merit of their own.

Several promising models demonstrate these principles in action:

Youth-led, adult-supported organizations maintain clear youth leadership while incorporating adult allies in supportive roles. Organizations like Zero Hour and Sunrise Movement exemplify this approach—founded and led by young people while strategically incorporating older advisors, mentors, and supporters who enhance youth impact without controlling organizational direction. This model maintains authentic youth leadership while leveraging adult resources and experience.

Intergenerational coalitions bring youth-led and adult-led organizations together through formal partnerships maintaining respective organizational autonomy while coordinating strategy, communications, and actions for greater combined impact. These arrangements respect distinct organizational identities and leadership while creating structured collaboration around shared goals. Groups like the Climate Action Network exemplify this coalition approach linking organizations across generational, geographic, and tactical differences.

Youth advisory bodies with substantive authority influence larger institutions through structured input mechanisms with real decision impact rather than symbolic consultation. When designed with authentic power-sharing rather than tokenism, these bodies enable youth perspective integration into institutions while maintaining institutional continuity and technical capacity. Examples include youth climate councils integrated into municipal sustainability departments with defined authority over specific decisions or resource allocations.

Intergenerational learning communities create ongoing knowledge exchange beyond single events or programs. These communities develop sustained relationships and mutual learning practices that deepen understanding over time rather than isolated interactions. Climate reality networks and citizen science initiatives often exemplify this approach through continuous engagement across age cohorts around shared interests and complementary skills.

These collaborative approaches recognize that the climate challenge requires mobilizing society's full capacities rather than fragmenting them along generational lines. Each generation brings distinctive strengths shaped by their historical position, life stage, and relationship to both climate change and the systems producing it. Leveraging these complementary strengths while respecting youth leadership creates more powerful responses than either youth-only or adult-dominated approaches could achieve independently.

Listening to the Future: What Youth Leadership Teaches Us

The youth climate movement doesn't just represent young people learning about climate change—it equally represents what older generations can learn from youth leadership. Rather than viewing youth activism merely as developmental experience preparing future leaders, this perspective recognizes that young people already lead in ways that offer valuable lessons for climate engagement across generations.

Several powerful lessons emerge from youth climate leadership:

Moral clarity creates transformative energy when not diluted by assumptions about political feasibility or system constraints. Youth movements demonstrate how direct ethical framing—speaking plainly about obligations to future generations and moral failure of climate inaction—mobilizes engagement more effectively than technical discussions about emissions percentages or policy mechanisms. This clarity doesn't reflect naive idealism but rather proper prioritization of fundamental values over short-term considerations. Older climate advocates sometimes lose this clarity through excessive accommodation to perceived political constraints, missing opportunities for narrative transformation that youth approaches demonstrate. It's like how children can see through elaborate adult justifications to underlying inconsistencies—sometimes apparent simplicity reflects clarity rather than naivete.

Digital organizing enables distributed leadership and rapid scaling beyond what traditional hierarchical structures achieve. Youth movements leverage digital platforms for coordination without centralized control, enabling impressive mobilization with minimal infrastructure. This approach distributes leadership broadly, allowing simultaneous local adaptation and global coherence through shared frames rather than top-down direction. While traditional organizations invest substantial resources in physical infrastructure and formal leadership chains, youth movements demonstrate how digital networks enable alternative organizational forms achieving remarkable impact with limited resources. It's like comparing distributed computing to mainframes—different architectural principles enabling different capabilities and resilience patterns (Earth.Org, 2023).

Integrative framing connects climate with broader concerns rather than treating it as isolated environmental issue. Youth movements consistently situate climate within intersecting questions about economic systems, social justice, racial equity, and democratic governance rather than siloing it as narrow technical problem. This approach appeals to diverse constituencies beyond traditional environmental identities by connecting climate with other priority concerns while addressing the challenge's actual interconnected nature. It demonstrates that building diverse climate movements requires moving beyond narrow environmental framing toward integrative approaches acknowledging the challenge's systemic dimensions. It's like understanding that human health requires considering interconnections between physical, mental, social, and environmental factors rather than treating each dimension in isolation.

Creative tactics break through attention barriers where conventional approaches fail to penetrate media environments and public consciousness. Youth movements employ innovative protest formats, compelling visuals, strategic disruption, personal narratives, and artistic elements that generate disproportionate visibility and engagement compared to traditional advocacy methods. These approaches demonstrate how impact often depends less on quantitative resources than qualitative creativity in crafting memorable interventions that people actually notice amid information saturation. It's like how viral content can reach millions with minimal production budget while expensive conventional messaging disappears without impact—creativity often matters more than resource intensity in contemporary communication environments.

Authentic communication resonates beyond policy jargon that often characterizes established climate advocacy. Youth leaders typically speak directly, personally, and emotionally rather than exclusively through technical terminology and policy frameworks. This communication connects beyond policy audiences to broader publics through relatable language, personal stakes, and genuine emotional engagement rather than detached analysis. The resulting resonance demonstrates that effective climate communication requires connecting with values, identities, and emotions rather than operating exclusively through technical arguments. It's like how the most effective teachers make material personally relevant rather than presenting it as abstract information disconnected from students' lives and concerns.

For adult climate advocates, the question isn't whether youth deserve patronizing approval but rather what their distinctive approaches reveal about effective climate engagement that established methods often miss or underutilize. The answer includes moral clarity, digital organizing, integrative framing, creative tactics, authentic communication, transformational vision, and future-oriented timeframes that expand the climate movement's capabilities when incorporated alongside other essential elements. Like any effective team, climate progress requires leveraging diverse strengths rather than expecting uniform contributions from all participants regardless of their distinctive capacities and perspectives.

Conclusion: The Future Has a Voice

Today's youth climate leaders didn't choose to inherit a climate-changing world, but they have chosen to respond with remarkable moral clarity, strategic sophistication, and practical impact despite limited conventional resources and political power. Their movements demonstrate both what becomes possible when those most affected by climate change's future impacts fully engage with present decisions and what established approaches often miss through excessive accommodation to perceived constraints within existing systems.

The most important insight from youth climate leadership may be this: the future has a voice—and a remarkably articulate, determined voice at that. Rather than accepting that those who will experience climate impacts across their entire lives should wait silently while others with far shorter future stakes make decisions shaping those lives, youth activists have demanded active participation in decisions directly affecting their futures. This insistence on intergenerational voice represents not entitlement but appropriate democratic inclusion of those most affected by decisions with multi-generational consequences (Common Dreams, 2025).

This voice manifests not through uniform perspective but through diverse youth climate leadership reflecting different priorities, tactics, backgrounds, and emphasizes across geography, class, race, and other dimensions. From indigenous youth defending ancestral territories against extraction to urban students advocating green infrastructure in underserved neighborhoods, from student divestment campaigners transforming institutional investment to young entrepreneurs developing climate solutions, youth climate engagement spans a rich spectrum unified more by generational stake than specific approach.

What these diverse leaders share is recognition that climate change fundamentally affects their future possibilities—not as abstract concern but as concrete reality they will navigate throughout their lifetimes regardless of career paths or other life choices. This unavoidable relationship with climate impacts creates both motivation and perspective distinct from those with shorter future horizons. It doesn't make youth perspectives automatically superior on all climate dimensions, but it does make them essential voices in developing responses commensurate with the challenge's intergenerational nature.

The continued evolution of youth climate leadership faces several key questions:

Movement institutionalization tensions between maintaining spontaneous energy and developing sustainable structures for long-term impact. How youth climate initiatives navigate formalization, resource development, leadership transitions, and partnerships while preserving distinctive characteristics will significantly shape their future influence.

Diversification challenges regarding whose youth perspectives shape movement priorities and public perception. How movements incorporate leadership from diverse backgrounds, center frontline community perspectives, address internal equity considerations, and develop accountability mechanisms will determine both their justice impacts and resilience against fragmentation.

Sectoral expansion beyond protest and advocacy toward implementation roles in various climate solutions. How youth leadership extends from demand-side pressure toward supply-side participation in designing, implementing, and scaling climate solutions across sectors will influence both movement sustainability and overall climate progress.

Intergenerational collaboration approaches that maintain youth leadership while leveraging complementary adult capacities. How movements develop partnership models that preserve authentic youth direction while incorporating technical expertise, institutional access, and resource support will affect their ability to translate moral clarity into durable system change.

Personal sustainability amid intense commitment and emotional demands. How young activists balance climate engagement with personal wellbeing, educational requirements, relationship development, and other dimensions of healthy development will determine both individual flourishing and movement continuity beyond exceptional commitment periods.

What remains certain is that youth climate leadership isn't a temporary phenomenon but an essential element of effective climate response. Those who will experience climate impacts and transitions throughout their entire lives deserve substantive voice in decisions shaping those impacts and transitions. This voice contributes not just moral legitimacy but practical effectiveness through insights, approaches, and capacities that complement other essential elements of comprehensive climate action.

The youth climate movement's most profound legacy may ultimately be changing not just specific policies but our understanding of who legitimately participates in decisions with intergenerational consequences. By insisting that those with greatest future stake deserve present voice, youth climate leaders challenge deeply embedded assumptions about decision authority while demonstrating practical alternatives through their own sophisticated engagement despite conventional power limitations.

The future has a voice—and if we're wise, we'll listen. Not because youth perspectives automatically trump other considerations, but because effective climate responses require the full range of insights, capacities, and moral clarity available across generations. The climate challenge demands nothing less than our collective best efforts, leveraging complementary strengths toward shared purpose rather than fragmenting across generational lines. Youth leadership has proven its essential value in this collective effort. The question now is whether the rest of society proves equally willing to contribute its distinctive capacities toward their shared future.

Chapter 15: So What Can I Actually Do?

Practical Pathways Toward a Habitable Planet

You've made it through fourteen chapters of climate science, impacts, and solutions. You've waded through carbon cycles, energy transitions, psychological barriers, youth movements, and complex systems—all with more charts than your gym teacher's clipboard. If your brain feels a bit like Earth's atmosphere—overloaded with new inputs and struggling to reach equilibrium—you're not alone.

At this point, you might be thinking: "Okay, I get it—climate change is real, it's serious, and we need to address it. But what am I actually supposed to DO about it?" It's a fair question. Understanding the problem is only valuable if it leads to effective action. And in a challenge as vast and complex as climate change, knowing where to start can feel overwhelming.

The good news is that meaningful climate action exists at every scale—from individual choices to system transformations—and at every level of commitment from casual to dedicated activism. This chapter offers practical pathways organized by both sphere of influence (personal, community, workplace, political) and level of engagement (starting points, next steps, deeper commitment). Think of it as a climate action menu where you can order based on your appetite for change and available resources.

But first, a crucial frame: individual actions matter not just for their direct emissions reductions (though these do add up collectively), but for the market signals they send, the conversations they start, the norms they shift, and the momentum they build toward broader systemic change. It's not "either personal action OR policy change"—it's both/and, with each reinforcing the other. We need citizens, consumers, community members, and political actors all pushing in the same direction. The climate challenge is too big and too urgent for false dichotomies about where change should begin (Vandenbergh and Gilligan 2017).

So let's get practical. What can you, personally, actually do to help address climate change? Let's explore your options.

Personal Sphere: The Power of One (Multiplied by Billions)

Your personal choices create direct impact through emissions reductions, while simultaneously strengthening demand for climate-friendly options and normalizing sustainable behaviors within your social networks. The most effective personal actions target high-impact areas—particularly transportation, home energy, food, and major purchases (Wynes and Nicholas 2017).

Starting Points: Low-Hanging Carbon Fruit

If you're just beginning your climate journey, these starting points create meaningful impact with minimal disruption to your current lifestyle:

Electrify your next vehicle purchase. When it's time to replace your current vehicle, choose an electric or plug-in hybrid option if feasible for your circumstances. EVs now offer comparable total cost of ownership to conventional vehicles in many markets when considering purchase incentives, lower operating costs, and reduced maintenance. The climate impact grows as electricity gets cleaner, creating a transportation option that automatically improves its carbon footprint over time. It's like buying a self-improving appliance (IPCC 2022).

Shift toward plant-forward eating. You don't need to become fully vegetarian or vegan to make a difference—even reducing meat consumption by one day per week creates meaningful impact, especially when reducing beef specifically. Try instituting "Meatless Monday" in your household, exploring plant-based alternatives to favorite dishes, or adopting a "reducetarian" approach that mindfully decreases animal product consumption without eliminating it entirely. The climate benefits are substantial, and your arteries might thank you too. It's like dating vegetables instead of immediately proposing marriage to them— start with one night a week and see how it goes (Willett et al. 2019).

Switch to renewable electricity. Many utilities offer options to source your electricity from renewable generation through green pricing programs, community solar participation, or competitive suppliers. These programs typically add modest cost to electricity bills while creating significant climate benefit. If these options aren't available, residential solar installation may be viable depending on your location, roof characteristics, and financial situation. The investment often pays for itself while creating decades of clean energy generation (IPCC 2022).

Optimize home energy use. Conduct a DIY energy audit to identify and address the biggest energy wasters in your home. Common culprits include inadequate insulation, air leaks around windows and doors, inefficient lighting, and energy-hogging appliances. Simple fixes like weather stripping, LED bulbs, smart power strips, and programmable thermostats often pay for themselves quickly through utility bill savings while reducing emissions. It's like finding money in your couch cushions that also helps save the planet (EPA 2023).

Reconsider air travel. Flying generates significant emissions, particularly for long-distance trips. Consider whether some trips could be replaced with virtual connections, alternative transportation modes, or closer destinations. When flying remains necessary, direct flights at full capacity create fewer emissions than connecting flights or half-empty planes. Some airlines and third-party programs offer carbon offset options, though their quality varies significantly. The most impactful approach combines reduced flight frequency with carefully selected high-quality offsets when flying remains necessary. After all, even Leonardo DiCaprio occasionally needs to look at his private jet and ask, "Will there be room for both of us on this door, or should I take the train?" (Wynes and Nicholas 2017).

Each of these starting points creates meaningful impact with relatively minor lifestyle adjustments. They represent the low-hanging fruit of personal climate action—the highest impact-to-effort ratio for those beginning their climate journey.

Next Steps: Intermediate Impact Areas

Ready for more substantial engagement? These next steps deepen your climate impact through more significant lifestyle adjustments:

Drive less when alternatives exist. While electrifying your vehicle reduces emissions per mile, driving less eliminates those emissions entirely while providing health benefits through more active transportation. Experiment with walking, cycling, public transit, carpooling, or trip consolidation for some regular trips. Start small—perhaps replacing one car trip per week—and gradually increase as you develop comfort with alternatives. Many people discover they enjoy alternative transportation once they overcome initial hesitation, finding unexpected benefits in stress reduction, exercise, community connection, and time for podcasts or reading on transit. It's like discovering that legs were a transportation option all along! (Brand et al. 2021).

Adopt a climate-friendly diet. Beyond occasional plant-based meals, consider a more comprehensive dietary shift toward climate-friendly eating. This typically means emphasizing plant proteins (legumes, nuts, seeds), choosing lower-impact animal products when consumed (poultry instead of beef, mussels instead of shrimp), prioritizing whole foods over highly processed options, and reducing food waste through better planning, storage, and creative leftover usage. The specific approach varies with individual preferences, health needs, and cultural traditions— there's no one-size-fits-all climate diet. Remember, even cows started as vegetarians, and look what happened to their emissions (Poore and Nemecek 2018).

Weatherize your home. Beyond simple energy efficiency measures, comprehensive weatherization involves more substantial improvements like adding insulation, upgrading windows, sealing ductwork, and addressing major air leakage pathways. These investments typically reduce energy consumption 15-30% while improving comfort and reducing outdoor noise. Various incentive programs, tax credits, and financing options make weatherization more affordable, often with particular support for lower-income households. It's like giving your home a cozy, energy-efficient sweater that pays for itself (DOE 2023).

Embrace circularity in consumption. Reduce overall material consumption and its associated emissions by embracing circular economy principles in your purchasing. Buy durable, repairable items rather than disposable ones. Choose second-hand, refurbished, or upcycled products when appropriate. Repair rather than replace when items break. Share infrequently used items through borrowing networks or rental services rather than purchasing your own. These approaches typically save money while reducing emissions from manufacturing, transportation, and disposal. Marie Kondo your carbon footprint by asking, "Does this purchase spark climate joy?" (Ellen MacArthur Foundation 2021).

Shift financial services. Move your banking, investment, and insurance relationships toward institutions with stronger climate commitments. Major financial institutions use your deposits, premiums, and investments to finance various activities—including fossil fuel development for many mainstream banks. Research alternative providers with explicit policies against financing fossil fuel expansion and positive commitments toward climate solutions. These shifts create pressure within financial systems while aligning your money with your values. After all, your savings account shouldn't be secretly funding its own underwater future (Rainforest Action Network 2023).

These next steps typically require more significant lifestyle adjustments than starting points, but remain accessible for many people without extraordinary sacrifice or resource requirements. They deepen your climate impact while often providing co-benefits like health improvements, cost savings, and increased resilience to climate disruptions.

Deeper Commitment: Comprehensive Personal Climate Action

For those ready for deeper engagement, these approaches integrate climate considerations throughout major life decisions:

Electrify everything at home. When replacing home systems and appliances, systematically convert from fossil fuels to efficient electric alternatives. Replace gas furnaces with heat pumps, gas water heaters with heat pump or resistance models, gas stoves with induction, and gas dryers with heat pump versions. Combined with renewable electricity, this transition creates a fully decarbonized home while often improving performance. The transition can happen gradually as equipment reaches end-of-life, rather than requiring simultaneous replacement. It's like breaking up with fossil fuels one appliance at a time (Rewiring America 2023).

Transform your transportation. Create a comprehensive low-carbon mobility system for your household by combining electric vehicles, active transportation, public transit, carpooling, car-sharing, and trip reduction. The specific mix varies with location and circumstances, but the goal remains minimizing transportation emissions through systematic changes rather than isolated adjustments. This transformation often requires rethinking commuting patterns, shopping habits, recreation choices, and even housing location over time. Being a transportation polyamorist means having relationships with many mobility options rather than remaining monogamous with your car (IPCC 2022).

Adopt "climate-conscious" decision filters. Integrate climate considerations explicitly into all major life decisions from housing location to job selection to family planning to recreation choices. This doesn't mean climate becomes the only factor, but rather that you consciously evaluate climate implications alongside other priorities when making significant choices. Over time, these deliberate choices compound to create a lifestyle with dramatically lower emissions and greater resilience to climate impacts. It's like having a little Greta Thunberg sitting on your shoulder asking, "How dare you?" before every major purchase (Clayton et al. 2023).

Develop climate literacy. Invest in ongoing climate education beyond casual information consumption. Take online courses, read books (like this one!), attend lectures, and follow reputable climate scientists and communicators to deepen your understanding of both problems and solutions. This knowledge investment enables more effective decision-making, better resource allocation, and more persuasive communication with others about climate issues. It's like learning a new language that helps you navigate an increasingly climate-disrupted world (UNESCO 2022).

Create a personal carbon budget. Calculate your carbon footprint using reputable online tools, then establish reduction targets and track progress over time. This systematic approach enables data-driven decisions about where your emissions reduction efforts will have greatest impact. The process resembles financial budgeting—analyzing where your "carbon dollars" go, identifying areas for reduction, and measuring progress toward goals. This quantitative framework helps prioritize efforts where they'll make most difference rather than focusing exclusively on visible but potentially lower-impact actions. It's carbon Weight Watchers—everything gets measured and no action is too small to count (Carbon Trust 2023).

These deeper commitments represent substantial lifestyle integration of climate considerations—not occasional actions but comprehensive approaches that influence many aspects of daily life. They typically develop over time through incremental changes rather than overnight transformation, gradually creating lifestyles aligned with climate stability.

Importantly, personal climate actions create ripple effects beyond direct emissions reductions. When you install solar panels, your neighbors become more likely to consider them too. When you discuss your positive experiences with plant-based meals or electric vehicles, you reduce perceived barriers for others considering similar choices. When you demonstrate satisfaction with climate-friendly lifestyle adjustments, you challenge narratives that climate action requires sacrifice or deprivation. These social influences often exceed the direct emissions impact of individual actions through norm-shifting and social diffusion (Bollinger and Gillingham 2012).

Community Sphere: Collective Power

While individual actions matter, many climate solutions work better—or only work—at community scale. Community engagement leverages collective resources, expands impact beyond individual capacity, and builds social networks that sustain longer-term climate commitment (Ebi et al. 2018).

Starting Points: Entry-Level Community Engagement

These starting points require minimal time commitment while building connections with climate-concerned neighbors:

Join local climate or environmental groups. Most communities have organizations working on climate issues—from established environmental nonprofits to newer climate-specific groups to faith-based environmental initiatives. Attending meetings introduces you to like-minded neighbors while providing information about locally relevant climate issues and actions. These connections reduce the isolation many feel around climate concerns while creating pathways to more substantial engagement as your interest develops. It's like climate dating—you don't have to marry the first group you meet, but you should definitely start seeing people (Corner et al. 2020).

Participate in community clean-up or planting events. Many communities organize regular activities to clean parks, waterways, and public spaces or plant trees and gardens. These events provide immediate environmental benefits while building social connections around shared environmental values. They typically require just a few hours of commitment with no specialized knowledge, making them accessible entry points for community climate engagement. Plus, nothing builds camaraderie like collectively finding mystery objects in the local creek (Kardan et al. 2015).

Support local climate-friendly businesses. Research and patronize businesses in your community that demonstrate climate leadership through renewable energy installation, energy efficiency investments, sustainable transportation, waste reduction, or other climate-friendly practices. Many directories and certification programs help identify these businesses. Your spending creates direct financial support while demonstrating market demand for climate-conscious business practices. Vote with your wallet—one of the few voting systems where money literally buys influence, and for once, that's a good thing (Nielsen 2018).

Attend public meetings on climate-relevant issues. Local governments frequently hold public meetings on issues with climate implications—from transportation planning to building codes to waste management to energy policy. Simply attending these meetings signals public interest in climate considerations while providing information about local decision processes and priorities. Even without speaking, your presence as "climate constituency" influences official perception of public concern. Besides, local government meetings need more attendees under 80 who aren't angry about parking (Wang 2018).

Talk about climate. Engage friends, family, neighbors, and colleagues in constructive conversations about climate change—not confrontational arguments but authentic exchanges about concerns, experiences, and potential solutions. Research shows these conversations significantly influence climate attitudes and engagement, particularly when they focus on personal experiences, locally relevant impacts, and available solutions rather than abstract global statistics. You don't need expert knowledge—just willingness to share your authentic perspective and listen respectfully to others. Climate change should be like Fight Club—the first rule is that we absolutely DO talk about it (Geiger et al. 2021).

These community starting points require minimal commitment while connecting you with broader climate efforts. They lay foundation for deeper engagement while immediately expanding your impact beyond individual actions.

Next Steps: Collaborative Climate Projects

Ready for more substantive community engagement? These next steps involve collaborative projects addressing specific climate issues:

Start or join a community garden. Community gardens reduce food transportation emissions, sequester carbon in soil, provide environmental education opportunities, strengthen community resilience, and create green space with multiple benefits. Existing gardens typically welcome new participants, while starting new gardens usually requires identifying suitable space, recruiting interested neighbors, and navigating local regulations. The specific approach varies with local circumstances, but community gardens offer accessible entry points combining immediate action with long-term capacity building. Plus, nothing builds community like collective confusion over whether that's a weed or the heirloom tomato someone planted (Gough and Accordino 2013).

Organize community efficiency initiatives. Collaborative approaches to energy efficiency often achieve better results than individual efforts alone. Options include bulk purchasing programs that negotiate discounts on solar panels, heat pumps, or insulation; energy efficiency challenges where neighborhoods compete for greatest reduction; and team weatherization events where participants help improve each other's homes. These approaches reduce costs through economies of scale while creating social reinforcement for efficiency improvements. Think of it as an energy savings pyramid scheme where everyone actually wins (DOE 2022).

Develop community emergency response plans. Climate change increases frequency and intensity of extreme weather events affecting many communities. Collaborative emergency planning builds resilience through identifying vulnerable community members, establishing communication systems, designating emergency shelters, creating supply caches, and practicing response protocols. These preparations serve immediate disaster resilience while building community relationships valuable for longer-term climate action. When climate chaos comes calling, you'll want more of a plan than "hope my neighbor with the generator likes me" (FEMA 2023).

Create sharing systems to reduce consumption. Tool libraries, toy lending collections, community workshops, seed exchanges, and similar sharing systems reduce resource consumption through collaborative ownership models. These approaches recognize that many items spend most time unused in individual ownership, creating opportunities for significant resource efficiency through sharing rather than duplicative purchasing. While specific models vary, most sharing systems create both environmental and social benefits through reduced consumption and increased community connection. The average drill is used for just 12-15 minutes in its entire lifetime—making it the feline of the tool world: mostly sleeping but occasionally intensely productive (Frenken and Schor 2017).

Advocate for low-carbon transportation options. Collective advocacy can significantly improve conditions for walking, cycling, public transit, and electric vehicle adoption in your community. Specific approaches include attending transportation planning meetings, organizing neighborhood walkability audits, advocating for protected bike infrastructure, supporting public transit funding, or promoting EV charging installation. These efforts strengthen sustainable transportation systems that benefit entire communities rather than just motivated individuals. After all, even the most committed cyclists appreciate not having to choose between certain death and more certain death when navigating your average American stroad (Plumer and Popovich 2019).

These collaborative climate projects require more significant commitment than initial community engagement but create correspondingly greater impact. They typically develop specific skills while building relationships that support sustained climate action beyond individual efforts.

Deeper Commitment: Transformative Community Initiatives

These approaches pursue fundamental community transformation toward sustainability and resilience:

Develop community solar projects. Community solar programs enable multiple participants to share benefits from a single solar installation, making renewable energy accessible to those who can't install their own systems due to renting, unsuitable roofs, upfront cost barriers, or other constraints. While specific models vary from cooperative ownership to subscription programs, all expand clean energy access beyond affluent homeowners. Developing these projects typically requires navigating technical, financial, and regulatory complexities, but creates substantial impact through democratizing renewable energy access. It's solar panels for the people, by the people— a renewable revolution where everyone's invited (NREL 2022).

Create climate victory gardens. Inspired by historical "victory gardens" during wartime resource constraints, climate victory gardens apply regenerative agriculture practices at neighborhood scale to sequester carbon, build climate resilience, and reduce food system emissions. These initiatives combine food production with carbon-sequestering practices like no-till cultivation, cover cropping, compost application, and perennial integration. They serve both mitigation and adaptation functions while building community food security during climate disruption. Fight climate change with the power of zucchini overproduction! (Okvat and Zautra 2011).

Establish repair cafés and circular economy hubs. These community spaces combine tool access, skill sharing, and designated events focused on repairing rather than replacing broken items. They combat planned obsolescence and throwaway culture through practical skill development, waste reduction, and social reinforcement for circular economy principles. Beyond environmental benefits, they preserve traditional repair skills while building community resilience less dependent on fragile global supply chains. In a world of "I'll just buy a new one," channel your inner grandparent who would say "We're not made of money! This can be fixed!" (Charter and Keiller 2014).

Lead community climate planning processes. Comprehensive community climate action plans integrate mitigation and adaptation across sectors including energy, transportation, buildings, waste, and land use. While local governments often lead these processes, community initiatives can catalyze, inform, or complement official planning through neighborhood climate councils, community vulnerability assessments, citizen climate assemblies, or similar approaches. These participatory processes ensure climate plans reflect community priorities, leverage local knowledge, and maintain implementation momentum beyond electoral cycles. Because saving the world works better with a decent roadmap (Romero-Lankao et al. 2018).

These transformative community initiatives represent deep engagement requiring substantial time commitment, technical knowledge development, and multi-stakeholder collaboration. They create impact far beyond individual capacity while building durable systems and relationships supporting ongoing climate action across multiple domains.

The community sphere offers particular value in developing social infrastructure alongside physical changes—building relationships, skills, and cooperative capacity essential for climate resilience. When climate impacts intensify, these community connections often prove as important as technical solutions in determining vulnerability and response effectiveness. The social capital developed through community climate initiatives serves multiple purposes beyond direct emissions reduction or physical adaptation.

Workplace Sphere: Professional Climate Impact

Most adults spend substantial time in workplace settings that present distinctive opportunities for climate action beyond personal and community spheres. Workplace climate engagement leverages professional skills, institutional resources, and organizational networks toward substantial impact that individuals alone cannot achieve (Burger et al. 2020).

Starting Points: Climate Leadership From Any Position

These approaches require minimal organizational authority while demonstrating climate commitment in professional contexts:

Form or join a green team. Many organizations have employee-led sustainability committees addressing workplace environmental impacts through education, operational improvements, and management engagement. These teams typically work alongside formal job responsibilities rather than replacing them. Participation develops professional relationships around shared environmental values while creating structured pathways for workplace climate action regardless of official organizational commitment. It's the workplace equivalent of starting a book club, except this one might actually save the world (Robertson and Barling 2015).

Reduce work travel emissions. Business travel often represents a significant portion of workplace carbon footprints, particularly in professional service organizations. Advocating for thoughtful travel policies—prioritizing virtual connection when effective, choosing lower-carbon transportation when travel remains necessary, selecting efficient accommodation, and compensating unavoidable emissions through high-quality offsets—creates tangible emissions reductions while demonstrating climate leadership. The COVID-19 pandemic demonstrated remote collaboration effectiveness for many functions previously assumed to require in-person interaction. The bonus: fewer uncomfortable hotel mattresses and continental breakfasts featuring suspiciously rubbery eggs (Klöwer et al. 2020).

Promote sustainable commuting. Transportation to and from work represents a major emissions source for many organizations. Employee advocacy can strengthen sustainable commuting through improved bicycle facilities, preferential carpool parking, transit subsidies, EV charging installation, and flexible work arrangements reducing commute frequency. These approaches typically align with employee quality-of-life improvement while reducing organizational carbon footprint and transportation costs. Fewer cars in the parking lot means more space for actual humans—what a concept! (Ralph et al. 2020).

Reduce workplace waste. Many workplaces generate substantial waste through single-use items, inefficient printing practices, disposable food service products, and poor recycling systems. Employee initiatives addressing these issues through improved signage, education campaigns, sustainable procurement policies, and operational changes can significantly reduce waste while often saving money. These visible sustainability improvements demonstrate environmental commitment while creating pathways toward more comprehensive workplace climate action. After all, nothing says "I care about the future" like a passive-aggressive note about printing double-sided (EPA 2022).

Share climate knowledge professionally. Integrate climate awareness into your professional interactions through brown-bag presentations, knowledge-sharing in meetings, relevant article circulation, and discussion of climate implications for your industry or profession. These approaches raise climate awareness in professional contexts without requiring official organizational endorsement or authority. The specific content varies with professional context, but most fields have climate dimensions worth exploring with colleagues. Become the person who makes climate relevant to everyone's job—without becoming the person everyone avoids in the break room (Anderson 2021).

These workplace starting points require minimal institutional authority while demonstrating professional climate leadership. They create foundation for more substantial workplace engagement while developing relationships with similarly concerned colleagues.

Next Steps: Influencing Organizational Direction

These approaches actively shape organizational practices and priorities around climate considerations:

Advocate for energy efficiency improvements. Most organizations waste energy through inefficient equipment, poor controls, behavioral patterns, and operational practices. Employee advocacy for efficiency improvements—from lighting upgrades to HVAC optimization to smart controls to building envelope improvements—often succeeds because these investments typically deliver positive financial returns alongside emissions reductions. Framing efficiency around cost savings rather than exclusively environmental benefits often increases acceptance in budget-conscious organizations. Few executives can resist the seductive whisper of "reduced operating costs" (Stuart et al. 2020).

Promote sustainable procurement. Organizational purchasing decisions significantly influence carbon footprint and broader environmental impact. Employee advocacy for sustainable procurement can include developing environmental criteria for vendor selection, specifying low-carbon alternatives when available, prioritizing circular economy principles like repairability and recyclability, and considering embodied carbon alongside operational characteristics. These approaches leverage organizational purchasing power toward systemic change beyond what individual consumers can achieve. When the office orders 10,000 pens, it actually matters which pens they choose (CDP 2023).

Encourage renewable energy adoption. Organizations typically have greater capacity than individuals to implement renewable energy through on-site generation, power purchase agreements, renewable energy certificates, or community solar participation. Employee advocacy for renewable energy adoption—particularly when framed around cost stability, resilience benefits, and corporate reputation alongside environmental considerations—can significantly influence organizational energy decisions toward lower-carbon alternatives. Help your organization see that "powered by renewable energy" looks better on the website than "powered by whatever's cheapest, we don't really care" (IRENA 2022).

Support flexible work arrangements. Remote and hybrid work models can significantly reduce commuting emissions, office energy consumption, and physical space requirements when thoughtfully implemented. Employee advocacy for these arrangements—citing productivity research, talent attraction, space efficiency, and emissions reduction—can influence organizational policies toward more climate-friendly work models. While not appropriate for all functions, flexible arrangements offer significant emissions reduction potential alongside quality-of-life benefits for many knowledge workers. Less time commuting means more time for literally anything else (Breuer et al. 2021).

These approaches actively influence organizational direction rather than merely demonstrating individual commitment within workplace contexts. They typically require building coalitions with like-minded colleagues, developing business cases for proposed changes, and persistent advocacy through organizational decision processes.

Political Sphere: Systemic Change

While personal, community, and workplace actions create significant impact, addressing climate change at necessary scale and speed ultimately requires policy changes at all government levels. Political engagement leverages democratic systems toward structural changes enabling broader climate action beyond what individual choices alone can achieve (IPCC 2022).

Starting Points: Basic Climate Citizenship

These approaches represent foundational political engagement accessible to most people regardless of previous political experience:

Vote with climate in mind. Perhaps the most basic but essential climate political action involves considering candidates' climate positions when voting at all levels from local to national. This doesn't necessarily mean single-issue voting, but rather including climate as significant factor within your overall electoral decision-making. Research candidates' stated positions, voting records, and endorsements from environmental organizations to inform your choices. While individual votes may seem insignificant, collective electoral outcomes dramatically shape climate policy possibilities. Democracy: it's not just for complaining about afterward (Kotcher et al. 2018).

Contact elected representatives. Legislators at all levels pay attention to constituent communications, particularly on issues where they hear from relatively few people. Calls, emails, letters, and in-person visits expressing support for climate action help counter fossil fuel industry influence while demonstrating electoral importance of climate issues. These communications prove particularly effective when personalized rather than identical form messages, connected to local concerns rather than abstract global issues, and focused on specific legislation or decisions rather than general climate sentiment. Believe it or not, those form letters actually get counted, even if they don't get read (van Staalduinen et al. 2022).

Support climate ballot measures. Many jurisdictions allow direct democracy through ballot initiatives addressing specific policies. Climate-related measures might involve renewable energy standards, transportation funding, conservation programs, or other issues where voters directly decide policy rather than electing representatives who then make decisions. Research these measures when they appear on your ballot, considering their climate implications alongside other factors. These direct votes often determine consequential climate policies without intermediate political filtering. When it comes to climate, sometimes the best politician is no politician at all (Rogers 2020).

Attend public hearings. Government agencies at all levels hold public hearings on decisions with climate implications—from utility resource planning to transportation projects to building codes to land use decisions. Simply attending these hearings signals public interest in climate considerations, while providing testimony further strengthens this signal. Even if specific decisions don't directly reflect your input, consistent public presence around climate concerns gradually shifts official perception of constituent priorities. Public hearings: come for the democracy, stay for the weird cast of local characters who attend every single one (Wang 2018).

These basic civic engagement approaches require minimal time commitment while leveraging existing democratic systems toward climate-friendly outcomes. They connect individual voice to collective policy decisions while building foundation for more substantial political engagement.

Next Steps: Active Climate Advocacy

These approaches involve more active participation in policy development and advocacy:

Join advocacy organizations. Climate-focused advocacy groups leverage collective voice toward policy change through coordinated campaigns, lobbying, legal action, communications, and electoral work. Joining and supporting these organizations—through membership, donations, volunteer time, and participation in organized activities—amplifies your individual voice through collective action. Different organizations employ various theories of change, tactics, and issue priorities, enabling alignment with your specific concerns and preferred advocacy styles. Find your climate team—whether you're a street protest person or a wonky policy person, there's a group that needs exactly your energy (Martinez and McManus 2022).

Engage in public comment processes. Government agencies at all levels request public input on proposed regulations, plans, and decisions through formal comment processes. Submitting substantive comments addressing climate dimensions of these proposals—whether supporting climate-friendly elements or identifying concerns about climate-damaging aspects—creates official record that agencies must consider in their decisions. These comments prove most effective when addressing specific proposal elements with factual support rather than general climate sentiment. Public comments: where "I'm very angry about this" can become legal evidence if you phrase it correctly (Wagner et al. 2020).

Cultivate relationships with officials. Developing ongoing relationships with elected and appointed officials—through regular communication, meeting attendance, event participation, and constituency service interactions—creates foundation for more effective climate advocacy than isolated contacts. These relationships help officials see you as engaged constituent deserving attention rather than anonymous correspondent easily ignored. While building these relationships requires sustained effort, the resulting influence typically exceeds what isolated advocacy actions achieve. Remember, politicians are people too—weird, approval-seeking people who need to hear from you (Pielke 2018).

Participate in climate demonstrations. Public demonstrations supporting climate action—from local rallies to larger marches to coordinated global climate strikes—visibly display public concern while creating community around shared climate commitment. These events communicate climate urgency to both officials and broader public through media coverage, physical presence, and creative messaging. Participating demonstrates climate prioritization among constituents while connecting with others sharing similar concerns, creating both public pressure and community reinforcement. Plus, you get to exercise your clever sign-making skills (Fisher and Nasrin 2021).

Support climate champions. Beyond simply voting for climate-friendly candidates, active support through campaign volunteering, financial contributions, voter outreach, candidate forums, and public endorsements helps elect officials prioritizing climate action. These activities strengthen electoral prospects for climate champions while demonstrating political viability of ambitious climate platforms. They help counter fossil fuel industry political influence through grassroots engagement rather than financial resources. In politics, showing up is literally half the battle (Bonvillian and Sarma 2021).

These active advocacy approaches require more significant time commitment than basic civic engagement but create correspondingly greater policy influence. They help shape political landscape toward climate action through sustained engagement with democratic processes beyond occasional voting.

Putting It All Together: Your Personal Climate Action Portfolio

With so many potential climate actions across different spheres and commitment levels, developing your personal climate action portfolio requires thoughtful consideration of your specific circumstances, capacities, and priorities. Rather than attempting everything simultaneously or focusing exclusively on single approaches, effective climate engagement typically involves strategic combinations matching your situation.

Several principles help develop effective personal climate action portfolios:

Focus on high-impact areas first. Given limited time, energy, and resources, prioritizing actions with greatest climate impact per unit effort creates most effective engagement. While specific high-impact areas vary with individual circumstances, they typically include major emission sources like transportation, housing, food, and energy rather than visible but lower-impact issues like plastic straws or packaging that receive disproportionate attention. Strategic action requires accurate understanding of relative impact across different domains. Don't be the person obsessing over plastic straws while flying to three international conferences a year (Wynes and Nicholas 2017).

Consider your specific circumstances. Climate actions aren't one-size-fits-all—different approaches make sense for different people based on geographic location, housing situation, income, family responsibilities, professional context, and other factors. Someone in a walkable city with excellent public transit faces different transportation opportunities than someone in a rural area with minimal alternatives to personal vehicles. Effective climate engagement works with rather than against your specific circumstances, adapting general principles to particular contexts. The best climate action is the one you'll actually do, not the theoretically perfect one you'll perpetually feel guilty about not doing (Clayton et al. 2023).

Balance different spheres of influence. Combining actions across personal, community, workplace, and political spheres creates more comprehensive impact than focusing exclusively on any single domain. These spheres interact synergistically—personal actions build credibility for community leadership, political engagement creates conditions enabling better workplace outcomes, and so on. The specific balance varies with individual priorities and capacities, but integrating multiple spheres typically creates more robust engagement than isolated approaches. Be a climate portfolio diversification expert—your impact shouldn't depend on a single action any more than your retirement should depend on a single stock (Heller 2022).

Match commitment to available resources. Sustainable climate engagement aligns commitment levels with available time, energy, knowledge, and financial resources to prevent burnout from overcommitment or frustration from insufficient impact. Start with actions matching your current capacity while gradually building toward deeper engagement as circumstances allow. Remember that climate action represents marathon rather than sprint—sustained engagement over time typically creates greater impact than brief intensive effort followed by disengagement. Climate burnout helps no one—pace yourself for the long haul (Bamberg et al. 2023).

Leverage your specific strengths and passions. Most effective climate engagement builds on existing skills, knowledge, relationships, and interests rather than requiring entirely new capacities. An artist might create climate-themed works, a teacher might develop climate curriculum, an engineer might improve energy systems, a community organizer might build climate coalitions. These strength-based approaches feel more authentic and sustainable than generic actions disconnected from your identity and capabilities. The climate movement needs your unique superpowers, whatever they may be (Capstick et al. 2020).

Consider co-benefits alongside climate impact. Many climate actions provide significant benefits beyond emissions reduction—from health improvements through active transportation to cost savings through energy efficiency to community building through collaborative projects. Prioritizing actions with substantial co-benefits aligned with your other priorities creates more sustainable engagement by serving multiple values simultaneously rather than treating climate as isolated concern requiring sacrifice of other priorities. The best climate actions make your life better now while making everyone's life better later (Cohen et al. 2021).

Start somewhere and build gradually. Perhaps most importantly, begin your climate journey without waiting for perfect understanding or comprehensive plan. Start with actions matching your current knowledge, capacity, and circumstances while learning and evolving your approach over time. Like any significant life change, climate engagement develops through experience rather than emerging fully-formed from initial commitment. The most important step is the first one that begins your journey. Don't let the perfect be the enemy of the "actually getting started" (Markowitz and Shariff 2012).

Conclusion: The Climate Needs You

Throughout this book, we've explored climate science, impacts, and solutions across multiple domains. We've examined carbon cycles, energy systems, extreme weather, ocean changes, psychology, adaptation, youth leadership, and more—all to develop comprehensive understanding of both the climate challenge and potential responses. Now, in this final chapter, we've translated that understanding into practical action pathways across personal, community, workplace, and political spheres.

The climate needs you—your intelligence, creativity, relationships, resources, and persistent engagement—to help navigate one of the most complex challenges humanity has ever faced. It needs your personal choices reducing emissions and building momentum for broader change. It needs your community engagement developing local solutions and resilience. It needs your professional leadership integrating climate considerations throughout workplace contexts. And it needs your political voice supporting policies enabling systemic transformation.

While this collective challenge can feel overwhelming in its scale and complexity, remember that you're not alone. Millions of people worldwide are engaged in similar work across all domains—from scientists improving understanding to engineers designing solutions to activists building political will to community leaders developing local resilience. Together, these diverse contributions create climate action ecosystem far more powerful than any individual effort alone.

Your specific contribution matters not because it single-handedly solves the climate crisis, but because it represents essential thread in the larger tapestry of collective response. The question isn't whether your individual actions will fix everything—they won't—but rather what unique contributions you can make to humanity's most important collaborative project. When future generations look back at this pivotal moment in human history, what will they say about how you responded to the challenge you knew was coming?

Climate action requires both urgency and patience—urgency in implementing available solutions at maximum speed and scale, patience in recognizing that transformation takes time while continuing efforts despite setbacks and delays. It requires both individual commitment and collective mobilization, both technological innovation and social evolution, both practical pragmatism and moral vision. It requires your best contributions aligned with your specific capabilities, circumstances, and spheres of influence.

The path forward contains both peril and possibility—peril in potentially catastrophic consequences if we fail to address climate change adequately, possibility in the better world we might create through thoughtful response integrating climate solutions with other social priorities. This dual nature makes climate engagement both sobering responsibility and meaningful opportunity to participate in perhaps the most consequential transformation of human civilization since the industrial revolution itself.

So what will you actually do? The answer depends on your unique situation, but this chapter provides starting points across multiple domains matching different circumstances and commitment levels.

Whatever specific actions you choose, know that your engagement matters—not because any individual alone determines the outcome, but because our collective response emerges from millions of individual choices, commitments, and contributions like yours.

The climate needs you. Humanity needs you. Future generations need you.

The time for action is now.

Let's get to work.

Appendix A: Works Cited (by chapter)

Chapter 1 Works Cited

"Carbon Footprint Factsheet." Center for Sustainable Systems, University of Michigan, 2024, css.umich.edu/publications/factsheets/sustainability-indicators/carbon-footprint-factsheet.

"Climate Change." U.S. Environmental Protection Agency, 2024, www.epa.gov/climate-change.

"Climate Change 2021: The Physical Science Basis." Intergovernmental Panel on Climate Change, 2021, www.ipcc.ch/report/ar6/wg1/downloads/report/IPCC_AR6_WGI_SPM_final.pdf.

"Climate Change: Atmospheric Carbon Dioxide." NOAA Climate.gov, 9 Apr. 2024, www.climate.gov/news-features/understanding-climate/climate-change-atmospheric-carbon-dioxide.

"Climate Change Impacts on the Ocean and Marine Resources." U.S. Environmental Protection Agency, 19 Oct. 2022, www.epa.gov/climateimpacts/climate-change-impacts-ocean-and-marine-resources.

"Climate change mitigation: reducing emissions." European Environment Agency, 2024, www.eea.europa.eu/en/topics/in-depth/climate-change-mitigation-reducing-emissions.

"Climate change widespread, rapid, and intensifying – IPCC." Intergovernmental Panel on Climate Change, 9 Aug. 2021, www.ipcc.ch/2021/08/09/ar6-wg1-20210809-pr/.

"Choices made now are critical for the future of our ocean and cryosphere – IPCC." Intergovernmental Panel on Climate Change, 25 Sep. 2019, www.ipcc.ch/2019/09/25/srocc-press-release/.

"Control methane to slow global warming — fast." Nature, 2021, www.nature.com/articles/d41586-021-02287-y.

"Energy is at the heart of the solution to the climate challenge." Intergovernmental Panel on Climate Change, 31 Jul. 2020, www.ipcc.ch/2020/07/31/energy-climatechallenge/.

"Greenhouse Effect 101." Natural Resources Defense Council, 29 Jan. 2025, www.nrdc.org/stories/greenhouse-effect-101.

"How is climate change impacting the world's ocean." United Nations, 2024, www.un.org/en/climatechange/science/climate-issues/ocean-impacts.

"In-depth Q&A: The IPCC's sixth assessment on how to tackle climate change." Carbon Brief, 11 Apr. 2022, www.carbonbrief.org/in-depth-qa-the-ipccs-sixth-assessment-on-how-to-tackle-climate-change/.

"Main Greenhouse Gases." Center for Climate and Energy Solutions, 13 Nov. 2024, www.c2es.org/content/main-greenhouse-gases/.

"Overview of Greenhouse Gases." U.S. Environmental Protection Agency, 16 Jan. 2025, www.epa.gov/ghgemissions/overview-greenhouse-gases.

"Sea Level Change." Intergovernmental Panel on Climate Change, 2018, www.ipcc.ch/site/assets/uploads/2018/02/WG1AR5_Chapter13_FINAL.pdf.

"Special Report on the Ocean and Cryosphere in a Changing Climate." Intergovernmental Panel on Climate Change, 2019, www.ipcc.ch/srocc/.

"The IPCC Scenarios." World Ocean Review, 2024, worldoceanreview.com/en/wor-5/climate-change-threats-and-natural-hazards/climate-change-and-the-coasts/the-ipcc-scenarios/.

Chapter 2 Works Cited

"Chapter 11: Weather and Climate Extreme Events in a Changing Climate." IPCC, 2021, www.ipcc.ch/report/ar6/wg1/chapter/chapter-11/.

"Chapter 8: Water Cycle Changes." IPCC, 2021, www.ipcc.ch/report/ar6/wg1/chapter/chapter-8/.

"Climate Change and Extreme Weather Linked in U.N. Climate Report." Eos, 2021, https://eos.org/articles/climate-change-and-extreme-weather-linked-in-u-n-climate-report.

"Climate Change and Extreme Weather." NASA Science, 2024, https://science.nasa.gov/climate-change/extreme-weather/.

"Climate Change Widespread, Rapid, and Intensifying – IPCC." IPCC, 2021, www.ipcc.ch/2021/08/09/ar6-wg1-20210809-pr/.

"Climate attribution tools critical for understanding extreme events." NOAA, 2023, www.noaa.gov/news-release/climate-attribution-tools-critical-for-understanding-extreme-events.

"Explainer: What the new IPCC report says about extreme weather and climate change." Carbon Brief, 2022, www.carbonbrief.org/explainer-what-the-new-ipcc-report-says-about-extreme-weather-and-climate-change/.

"Chapter 4: Water | Climate Change 2022: Impacts, Adaptation and Vulnerability." IPCC, 2022, www.ipcc.ch/report/ar6/wg2/chapter/chapter-4/.

"Polar Vortex: How the Jet Stream and Climate Change Bring on Cold Snaps." Inside Climate News, 2018, https://insideclimatenews.org/news/02022018/cold-weather-polar-vortex-jet-stream-explained-global-warming-arctic-ice-climate-change/.

"Q&A: How is Arctic warming linked to the 'polar vortex' and other extreme weather?" Carbon Brief, 2019, www.carbonbrief.org/qa-how-is-arctic-warming-linked-to-polar-vortext-other-extreme-weather/.

"The Polar Jet Stream and Polar Vortex." MIT Climate Portal, 2022, https://climate.mit.edu/explainers/polar-jet-stream-and-polar-vortex.

"Think You Know the Polar Vortex? Think Again." NOVA PBS, 2019, www.pbs.org/wgbh/nova/article/think-you-know-polar-vortex-think-again/.

"Understanding the Arctic polar vortex." NOAA Climate.gov, 2021, www.climate.gov/news-features/understanding-climate/understanding-arctic-polar-vortex.

"The Water Cycle and Climate Change." NASA Earth Observatory, 2010, https://earthobservatory.nasa.gov/features/Water/page3.php.

"The Water Cycle and Climate Change." UCAR Center for Science Education, 2022, https://scied.ucar.edu/learning-zone/climate-change-impacts/water-cycle-climate-change.

"The water cycle is intensifying as the climate warms, IPCC report warns." The Conversation, 2021, https://theconversation.com/the-water-cycle-is-intensifying-as-the-climate-warms-ipcc-report-warns-that-means-more-intense-storms-and-flooding-165590.

"Weather vs. Climate." Climate at a Glance, 2022, https://climateataglance.com/climate-at-a-glance-weather-vs-climate/.

"Shifting Winds: How a wavier polar jet stream causes extreme weather events." Arctic Council, 2024, https://arctic-council.org/news/shifting-winds-how-a-wavier-polar-jet-stream-causes-extreme-weather-events/.

"NASA Scientific Visualization Studio | Water Cycle Extremes: Droughts and Pluvials." NASA, 2023, https://svs.gsfc.nasa.gov/5392.

"Climate Change and Extreme Weather." World Meteorological Organization, 2023, https://wmo.int/about-us/world-meteorological-day/wmd-2022/climate-change-and-extreme-weather.

"Polar Vortex." UC Davis, 2021, https://www.ucdavis.edu/climate/definitions/what-is-the-polar-vortex.

Chapter 3 Works Cited

"AR6 Synthesis Report: Climate Change 2023." IPCC, 2023, www.ipcc.ch/report/sixth-assessment-report-cycle/.

"Carbon Dioxide." NASA Vital Signs of the Planet, 15 Aug. 2023, climate.nasa.gov/vital-signs/carbon-dioxide/.

"Climate Change: Atmospheric Carbon Dioxide." NOAA Climate.gov, 9 Apr. 2024, www.climate.gov/news-features/understanding-climate/climate-change-atmospheric-carbon-dioxide.

"Climate Feedback Loops and Tipping Points." Center for Science Education, scied.ucar.edu/learning-zone/earth-system/climate-system/feedback-loops-tipping-points.

"Climate Science - Feedback Loops & Tipping Points." The Borrowed Earth Project, 12 Oct. 2020, borrowedearthproject.com/blog/climate-science-feedback-loops.

"During a year of extremes, carbon dioxide levels surge faster than ever." National Oceanic and Atmospheric Administration, www.noaa.gov/news-release/during-year-of-extremes-carbon-dioxide-levels-surge-faster-than-ever.

"15 Climate Feedback Loops and Examples." Earth How, 25 Sep. 2023, earthhow.com/climate-feedback-loops/.

"How Feedback Loops Are Making the Climate Crisis Worse." The Climate Reality Project, 8 Jan. 2024, www.climaterealityproject.org/blog/how-feedback-loops-are-making-climate-crisis-worse.

"In-depth Q&A: The IPCC's sixth assessment on how to tackle climate change." Carbon Brief, 11 Apr. 2022, www.carbonbrief.org/in-depth-qa-the-ipccs-sixth-assessment-on-how-to-tackle-climate-change/.

IPCC. "Summary for Policymakers." Global Warming of 1.5 °C, www.ipcc.ch/sr15/chapter/spm/.

"IPCC Climate Change Reports." NRDC, 6 Jan. 2025, www.nrdc.org/stories/ipcc-climate-change-reports-why-they-matter-everyone-planet.

"IPCC report: urgent climate action needed to halve emissions by 2030." World Economic Forum, Apr. 2022, www.weforum.org/stories/2022/04/ipcc-report-mitigation-climate-change/.

"IPCC Sixth Assessment Report." Wikipedia, 11 Mar. 2025, en.wikipedia.org/wiki/IPCC_Sixth_Assessment_Report.

Lan, X., Tans, P. and K.W. Thoning. "Trends in globally-averaged CO_2 determined from NOAA Global Monitoring Laboratory measurements." NOAA Global Monitoring Laboratory, gml.noaa.gov/ccgg/trends/global.html.

Mercator Research Institute on Global Commons and Climate Change. "Remaining Carbon Budget." MCC-Berlin, www.mcc-berlin.net/en/research/co2-budget.html.

NASA Scientific Visualization Studio. "Atmospheric CO_2 Trends." 6 Jan. 2025, svs.gsfc.nasa.gov/30556.

"Permafrost carbon feedbacks threaten global climate goals." Proceedings of the National Academy of Sciences, www.pnas.org/doi/10.1073/pnas.2100163118.

"Substantial reductions in non-CO2 greenhouse gas emissions reductions implied by IPCC estimates of the remaining carbon budget." Communications Earth & Environment, www.nature.com/articles/s43247-023-01168-8.

"The evidence is clear: the time for action is now. We can halve emissions by 2030." IPCC, 4 Apr. 2022, www.ipcc.ch/2022/04/04/ipcc-ar6-wgiii-pressrelease/.

"The IPCC just published its summary of 5 years of reports – here's what you need to know." Climate Champions, 21 Mar. 2023, climatechampions.unfccc.int/the-ipcc-just-published-its-summary-of-5-years-of-reports-heres-what-you-need-to-know/.

"Top Findings from the IPCC Climate Change Report 2023." World Resources Institute, www.wri.org/insights/2023-ipcc-ar6-synthesis-report-climate-change-findings.

"Will climate feedback loops push us past a 'point of no return'?" MIT Climate Portal, climate.mit.edu/ask-mit/will-climate-feedback-loops-push-us-past-point-no-return.

Chapter 4 Works Cited

EPA. (2022). "Climate Change Impacts on the Ocean and Marine Resources." Environmental Protection Agency. https://www.epa.gov/climateimpacts/climate-change-impacts-ocean-and-marine-resources

IPCC. (2019). "Special Report on the Ocean and Cryosphere in a Changing Climate." Intergovernmental Panel on Climate Change. https://www.ipcc.ch/srocc/

IPCC. (2021). "Climate change widespread, rapid, and intensifying." Intergovernmental Panel on Climate Change. https://www.ipcc.ch/2021/08/09/ar6-wg1-20210809-pr/

Met Office. (2024). "The Atlantic Meridional Overturning Circulation in a changing climate." https://www.metoffice.gov.uk/blog/2024/the-atlantic-meridional-overturning-circulation-in-a-changing-climate

NCEI. (2024). "Decades of Data on a Changing Atlantic Circulation." National Centers for Environmental Information. https://www.ncei.noaa.gov/news/decades-data-changing-atlantic-circulation

NOAA. (2024). "Ocean acidification." National Oceanic and Atmospheric Administration. https://www.noaa.gov/education/resource-collections/ocean-coasts/ocean-acidification

NOAA Fisheries. (2024). "Climate Change: Understanding The Impacts." National Oceanic and Atmospheric Administration. https://www.fisheries.noaa.gov/topic/climate-change/understanding-the-impacts

NOAA Sanctuaries. (2024). "Blue Carbon in Marine Protected Areas." Office of National Marine Sanctuaries. https://sanctuaries.noaa.gov/science/conservation/blue-carbon-in-marine-protected-areas-part-1.html

The Blue Carbon Initiative. (2024). "About Blue Carbon." https://www.thebluecarboninitiative.org/

Chapter 5 Works Cited

"800,000 Year record of CO2 Concentration." Global Climate Change Impacts in the United States 2009 Report Legacy site, U.S. Global Change Research Program. https://nca2009.globalchange.gov/800000-year-record-co2-concentration/index.html

"Carbon Isotopes." NOAA Earth System Research Laboratories. https://www.esrl.noaa.gov/gmd/education/isotopes/

"Climate Change 2021: The Physical Science Basis." IPCC Sixth Assessment Report, 2021. https://www.ipcc.ch/report/ar6/wg1/

"Climate Change: Atmospheric Carbon Dioxide." NOAA Climate.gov, 2024. https://www.climate.gov/news-features/understanding-climate/climate-change-atmospheric-carbon-dioxide

"Climate Change: How Do We Know?" NASA Global Climate Change: Vital Signs of the Planet. https://climate.nasa.gov/evidence/

"Climate Change Science." EPA. https://www.epa.gov/climate-change-science

"Climate Change." NASA Science. https://science.nasa.gov/climate-change/

"Climate data monitoring." NOAA. https://www.noaa.gov/education/resource-collections/climate/climate-data-monitoring

"Climate Model Ensembles." NOAA. https://www.climate.gov/maps-data/primer/climate-model-ensembles

"Climate Models." NASA Global Climate Change: Vital Signs of the Planet. https://climate.nasa.gov/news/2943/climate-change-modeling-101/

"Climate Models." NOAA. https://www.climate.gov/maps-data/primer/climate-models

"Coral Paleoclimate Research." NOAA. https://www.ncei.noaa.gov/products/paleoclimatology/coral

"Downscaling Climate Information." EPA. https://www.epa.gov/arc-x/downscaling-climate-information

"Explainer: How the rise and fall of CO2 levels influenced the ice ages." Carbon Brief, 2021. https://www.carbonbrief.org/explainer-how-the-rise-and-fall-of-co2-levels-influenced-the-ice-ages/

"Global warming 'unequivocally' human driven: IPCC." United Nations, 2021. https://www.un.org/en/delegate/global-warming-'unequivocally'-human-driven-ipcc

"How are satellites used to observe the ocean?" National Ocean Service, NOAA. https://oceanservice.noaa.gov/facts/satellites-ocean.html

"How tree rings tell time and climate history." NOAA Climate.gov, 2018. https://www.climate.gov/news-features/blogs/beyond-data/how-tree-rings-tell-time-and-climate-history

"Responding to Climate Change." NASA Global Climate Change: Vital Signs of the Planet. https://climate.nasa.gov/solutions/resources/

"Satellite Measurements of Warming in the Troposphere." NASA Earth Observatory. https://earthobservatory.nasa.gov/features/Trop

"Scientific Consensus: Earth's Climate is Warming." NASA Global Climate Change: Vital Signs of the Planet. https://climate.nasa.gov/scientific-consensus/

"The Atmosphere: Getting a Handle on Carbon Dioxide." NASA Earth Observatory. https://earthobservatory.nasa.gov/features/CarbonCycle

"The Last Time the Globe Warmed." NASA Earth Observatory. https://earthobservatory.nasa.gov/features/GlobalWarming/page3.php

"Tree rings provide snapshots of Earth's past climate." NASA Climate Change, 2017. https://climate.nasa.gov/news/2540/tree-rings-provide-snapshots-of-earths-past-climate/

"What do ice cores reveal about the past?" National Snow and Ice Data Center. https://nsidc.org/learn/ask-scientist/core-climate-history

"What is coral bleaching?" NOAA. https://oceanservice.noaa.gov/facts/coral_bleach.html

"What does the full body of evidence tell us about global warming?" Skeptical Science. https://skepticalscience.com/evidence-for-global-warming.htm

Chapter 6 Works Cited

Carbon Brief. (2022, April 11). Explainer: What the new IPCC report says about extreme weather and climate change. https://www.carbonbrief.org/explainer-what-the-new-ipcc-report-says-about-extreme-weather-and-climate-change/

Center for Climate and Energy Solutions. (2023, July 14). Heat Waves and Climate Change. https://www.c2es.org/content/heat-waves-and-climate-change/

Center for Climate and Energy Solutions. (2023, July 14). Wildfires and Climate Change. https://www.c2es.org/content/wildfires-and-climate-change/

EPA. (2016, July 1). Climate Change Indicators: Coastal Flooding. https://www.epa.gov/climate-indicators/climate-change-indicators-coastal-flooding

EPA. (2024, November 19). Extreme Precipitation. https://www.epa.gov/climatechange-science/extreme-precipitation

EPA. (2024, November 22). EPA Researchers are Providing Tools and Resources to Prepare Communities for Climate Change and Extreme Storms. https://www.epa.gov/sciencematters/epa-researchers-are-providing-tools-and-resources-prepare-communities-climate-change

EPA. (2025, January 15). Climate Change Indicators: Heavy Precipitation. https://www.epa.gov/climate-indicators/climate-change-indicators-heavy-precipitation

EPA. (2025, February 4). Climate Change Indicators: Wildfires. https://www.epa.gov/climate-indicators/climate-change-indicators-wildfires

IPCC. (2021). Chapter 11: Weather and Climate Extreme Events in a Changing Climate. In Climate Change 2021: The Physical Science Basis. Contribution of Working Group I to the Sixth Assessment Report of the Intergovernmental Panel on Climate Change. https://www.ipcc.ch/report/ar6/wg1/chapter/chapter-11/

NASA. (2024, October 23). The Effects of Climate Change. https://science.nasa.gov/climate-change/effects/

NASA. (2025, February 6). Wildfires and Climate Change. https://science.nasa.gov/wildfires-and-climate-change/

NEEF. (n.d.). Heat Waves, Heat Islands, and Your Health. https://www.neefusa.org/story/climate-change/heat-waves-heat-islands-and-your-health

NOAA. (2016). Scientists: Strong evidence that human-caused climate change intensified 2015 heat waves. https://www.noaa.gov/media-release/scientists-strong-evidence-human-caused-climate-change-intensified-2015-heat-waves

NOAA. (n.d.). Wildfire climate connection. https://www.noaa.gov/noaa-wildfire/wildfire-climate-connection

Vicedo-Cabrera, A.M., Scovronick, N., Sera, F. et al. (2021). The burden of heat-related mortality attributable to recent human-induced climate change. Nature Climate Change. https://www.nature.com/articles/s41558-021-01058-x

WHO. (2024, May 28). Heat and health. https://www.who.int/news-room/fact-sheets/detail/climate-change-heat-and-health

Chapter 7 Works Cited

Climate Champions. (2023). "The IPCC just published its summary of 5 years of reports – here's what you need to know." UNFCCC Climate Champions. https://climatechampions.unfccc.int/the-ipcc-just-published-its-summary-of-5-years-of-reports-heres-what-you-need-to-know/

Dahl, Erik J. "Naval Innovation: From Coal to Oil." Joint Force Quarterly, no. 27, 2000, pp. 50-56. ResearchGate, https://www.researchgate.net/publication/235048025_Naval_Innovation_From_Coal_to_Oil.

IEA. (2023). "Renewables 2023 – Executive Summary." International Energy Agency. https://www.iea.org/reports/renewables-2023/executive-summary

IEA. (2024). "Renewables 2024 – Executive Summary." International Energy Agency. https://www.iea.org/reports/renewables-2024/executive-summary

IPCC. (2022). Special Report: Renewable Energy Sources and Climate Change Mitigation. Intergovernmental Panel on Climate Change. https://www.globalchange.gov/reports/ipcc-special-report-renewable-energy-sources-and-climate-change-mitigation

IRENA. (2021). "Renewable Energy and Jobs – Annual Review 2021." International Renewable Energy Agency. https://www.irena.org/publications/2021/Oct/Renewable-Energy-and-Jobs-Annual-Review-2021

IRENA. (2023). "Renewable energy and jobs: Annual review 2023." International Renewable Energy Agency. https://www.irena.org/Publications/2023/Sep/Renewable-energy-and-jobs-Annual-review-2023

IRENA. (2024). "Energy and Jobs." International Renewable Energy Agency. https://www.irena.org/Energy-Transition/Socio-economic-impact/Energy-and-Jobs

IRENA. (2025). "Record-Breaking Annual Growth in Renewable Power Capacity." International Renewable Energy Agency. https://www.irena.org/

RMI. (2025). "Peaking: A Brief History of Select Energy Transitions." Rocky Mountain Institute. https://rmi.org/insight/how-past-energy-transitions-foretell-a-quicker-shift-away-from-fossil-fuels-today/

Smil, V. (2010). Energy Transitions: History, Requirements, Prospects. Praeger.

Smil, V. (2017). Energy and Civilization: A History. MIT Press.

The Conversation. (2024). "Muscle, wood, coal, oil: what earlier energy transitions tell us about renewables." https://theconversation.com/muscle-wood-coal-oil-what-earlier-energy-transitions-tell-us-about-renewables-213550

Visual Capitalist. (2022). "Visualizing the History of Energy Transitions." https://www.visualcapitalist.com/visualizing-the-history-of-energy-transitions/

World Economic Forum. (2022). "The world's energy transitions: a history told in infographics." https://www.weforum.org/stories/2022/04/visualizing-the-history-of-energy-transitions/

World Economic Forum. (2023). "Renewable energy jobs double to 13.7m in 10 years, IRENA finds." https://www.weforum.org/stories/2023/10/irena-renewable-energy-jobs/

Chapter 8 Works Cited

Carbon Brief. (2018). Mapped: How 'proxy' data reveals the climate of the Earth's distant past. Retrieved from https://interactive.carbonbrief.org/how-proxy-data-reveals-climate-of-earths-distant-past/

Colorado Sun. (2022, June 24). Tree rings show Colorado River Basin drought could get much worse. Retrieved from https://coloradosun.com/2022/06/24/tree-rings-drought-colorado-river-basin/

IPCC. (2013). Climate Change 2013: The Physical Science Basis. Contribution of Working Group I to the Fifth Assessment Report of the Intergovernmental Panel on Climate Change. Cambridge University Press, Cambridge, United Kingdom and New York, NY, USA.

IPCC. (2021). Climate Change 2021: The Physical Science Basis. Contribution of Working Group I to the Sixth Assessment Report of the Intergovernmental Panel on Climate Change. Cambridge University Press, Cambridge, United Kingdom and New York, NY, USA.

NASA. (2017, January 27). Tree rings provide snapshots of Earth's past climate. Climate Change: Vital Signs of the Planet. Retrieved from https://climate.nasa.gov/news/2540/tree-rings-provide-snapshots-of-earths-past-climate/

Nature. (2022). Tree-ring data set for dendroclimatic reconstructions and dendrochronological dating in European Russia. Scientific Data, 9(1), 236. Retrieved from https://www.nature.com/articles/s41597-022-01456-6

NOAA. (2018, November 29). How tree rings tell time and climate history. NOAA Climate.gov. Retrieved from https://www.climate.gov/news-features/blogs/beyond-data/how-tree-rings-tell-time-and-climate-history

NOAA Climate.gov. (2018). Dendrochronology - Trees: Recorders of Climate Change. Retrieved from https://www.climate.gov/teaching/resources/dendrochronology-trees-recorders-climate-change

RealClimate. (2015, February 19). The mystery of the offset chronologies: Tree rings and the volcanic record of the 1st millennium. Retrieved from https://www.realclimate.org/index.php/archives/2015/02/the-mystery-of-the-offset-chronologies-tree-rings-and-the-volcanic-record-of-the-1st-millennium/

RealClimate. (2021, August 9). A Tale of Two Hockey Sticks. Retrieved from https://www.realclimate.org/index.php/archives/2021/08/a-tale-of-two-hockey-sticks/

ScienceDirect. (2013). Testing the hypothesis of post-volcanic missing rings in temperature sensitive dendrochronological data. Dendrochronologia, 31(3), 216-222. Retrieved from https://www.sciencedirect.com/science/article/abs/pii/S1125786513000258

USGS. (2022). Tree rings reveal unmatched 2nd century drought in the Colorado River Basin. Retrieved from https://www.usgs.gov/publications/tree-rings-reveal-unmatched-2nd-century-drought-colorado-river-basin

University of Georgia Extension. (2018). What do tree rings tell us about megadroughts in the Western US? Climate and Agriculture in the Southeast. Retrieved from https://site.extension.uga.edu/climate/2018/08/what-do-tree-rings-tell-us-about-megadroughts-in-the-western-us/

Chapter 9 Works Cited

350.org. (2023). "Fossil Free: Divestment." Retrieved from https://350.org/fossil-free/

Ansar, A., Caldecott, B., & Tilbury, J. (2023). "Stranded assets and the fossil fuel divestment campaign." Smith School of Enterprise and the Environment, University of Oxford.

Banking on Climate Chaos. (2024). "Banking on Climate Chaos 2024." Retrieved from https://www.bankingonclimatechaos.org/

BloombergNEF. (2024). "Global Clean Energy Investment Jumps 17%, Hits $1.8 Trillion in 2023." Retrieved from https://about.bnef.com/blog/global-clean-energy-investment-jumps-17-hits-1-8-trillion-in-2023-according-to-bloombergnef-report/

Delmas, M. A., & Burbano, V. C. (2023). "The Drivers of Greenwashing." California Management Review.

European Commission. (2023). "Sustainable Finance Disclosure Regulation." Retrieved from https://finance.ec.europa.eu/sustainable-finance/disclosures/sustainability-related-disclosure-financial-services-sector_en

IPCC. (2022). "Special Report on Global Warming of 1.5°C." Intergovernmental Panel on Climate Change.

IRENA. (2023). "World Energy Transitions Outlook 2023." International Renewable Energy Agency. Retrieved from https://www.irena.org/Digital-Report/World-Energy-Transitions-Outlook-2023

Mercure, J.F., Pollitt, H., & Viñuales, J.E. (2023). "Macroeconomic impact of stranded fossil fuel assets." Nature Climate Change.

NGFS. (2023). "Guide on climate-related disclosure for central banks." Network for Greening the Financial System. Retrieved from https://www.ngfs.net/system/files/import/ngfs/medias/documents/ngfs_guide_on_climate-related_disclosure_for_central_banks_-_second_edition.pdf

NOAA. (2024). "Billion-Dollar Weather and Climate Disasters." National Oceanic and Atmospheric Administration. Retrieved from https://www.ncei.noaa.gov/access/billions/

Project Drawdown. (2023). "Financial Innovation." Retrieved from https://drawdown.org/solutions/financial-innovation

Rainforest Action Network. (2024). "Banking on Climate Chaos: Fossil Fuel Finance Report 2024." Retrieved from https://www.ran.org/press-releases/bocc2024/

Sierra Club. (2024). "Top 6 U.S. Banks Financed Fossil Fuels with $1.8 trillion Since the Paris Agreement." Retrieved from https://www.sierraclub.org/press-releases/2024/05/top-6-us-banks-financed-fossil-fuels-18-trillion-paris-agreement-chase-citi

TCFD. (2023). "Task Force on Climate-related Financial Disclosures." Retrieved from https://www.fsb-tcfd.org/

U.S. Department of Energy. (2023). "Vineyard Wind." Retrieved from https://www.energy.gov/articles/vineyard-wind-commercial-scale-offshore-wind-farm

U.S. Department of Energy. (2024). "National Community Solar Partnership." Retrieved from https://www.energy.gov/communitysolar/community-solar

World Bank. (2024). "Global Carbon Pricing Revenues Top a Record $100 Billion." Retrieved from https://www.worldbank.org/en/news/press-release/2024/05/21/global-carbon-pricing-revenues-top-a-record-100-billion

World Economic Forum. (2023). "Over $1 trillion invested in green energy in 2022." Retrieved from https://www.weforum.org/stories/2023/02/energy-transition-investment-record-global-energy-crisis-3-february/

Chapter 10 Works Cited

California Independent System Operator (CAISO). (2013). What the duck curve tells us about managing a green grid. https://www.caiso.com/documents/flexibleresourceshelprenewables_fastfacts.pdf

Center for American Progress. (2022). Advancing equity through grid modernization. https://www.americanprogress.org/article/advancing-equity-grid-modernization/

Department of Energy (DOE). (2017). Confronting the duck curve: How to address over-generation of solar energy. https://www.energy.gov/eere/articles/confronting-duck-curve-how-address-over-generation-solar-energy

Department of Energy (DOE). (2022). NREL study identifies the opportunities and challenges of achieving the U.S. transformational goal of 100% clean electricity by 2035. https://www.energy.gov/eere/articles/nrel-study-identifies-opportunities-and-challenges-achieving-us-transformational-goal

Department of Energy (DOE). (2023a). National transmission needs study. https://www.energy.gov/gdo/national-transmission-needs-study

Department of Energy (DOE). (2023b). Grid planning: What's at stake for communities. https://elpc.org/blog/grid-planning-whats-at-stake-for-communities/

Department of Energy (DOE). (2023c). Energy equity and environmental justice. https://www.energy.gov/eere/energy-equity-and-environmental-justice

Department of Energy (DOE). (2024). Energy accessibility. https://www.energy.gov/eere/energy-accessibility

Environmental and Energy Study Institute (EESI). (2019). Fact sheet: Energy storage. https://www.eesi.org/papers/view/energy-storage-2019

Environmental Law & Policy Center. (2022). Grid planning: What's at stake for communities. https://elpc.org/blog/grid-planning-whats-at-stake-for-communities/

International Hydropower Association. (2024). Pumped storage hydropower: Water batteries for solar and wind power. https://www.hydropower.org/factsheets/pumped-storage

National Renewable Energy Laboratory (NREL). (2018). Ten years of analyzing the duck chart: How an NREL discovery in 2008 is helping enable more solar on the grid today. https://www.nrel.gov/news/program/2018/10-years-duck-curve.html

National Renewable Energy Laboratory (NREL). (2021). North American renewable integration study highlights opportunities for a coordinated, continental low-carbon grid. https://www.nrel.gov/news/program/2021/north-american-renewable-integration-study-highlights-opportunities-for-a-coordinated-continental-low-carbon-grid.html

National Renewable Energy Laboratory (NREL). (2022). NREL outlines paths and challenges of reaching 100% clean electric grid by 2035. https://www.publicpower.org/periodical/article/nrel-outlines-paths-and-challenges-reaching-100-clean-electric-grid-2035

National Renewable Energy Laboratory (NREL). (2024a). National transmission planning study. https://www2.nrel.gov/grid/national-transmission-planning-study

National Renewable Energy Laboratory (NREL). (2024b). Transmission planning. https://www2.nrel.gov/grid/transmission-planning

Chapter 11 Works Cited

Center for Sustainable Systems. 2022. "Carbon Footprint Factsheet." University of Michigan. https://css.umich.edu/publications/factsheets/sustainability-indicators/carbon-footprint-factsheet

Citizens' Climate Lobby. 2024. "Building Political Will for Climate Solutions." https://citizensclimatelobby.org/

Climate Resolve. 2023. "Community-based Climate Solutions." https://climateresolve.org/

EPA (Environmental Protection Agency). 2024. "Assumptions and References for Household Carbon Footprint Calculator." https://www.epa.gov/ghgemissions/assumptions-and-references-household-carbon-footprint-calculator

Feldman, Lauren, and P. Sol Hart. 2021. "Building connections between climate change communication and community engagement." International Journal of Environmental Research and Public Health 18(13): 6932. https://doi.org/10.3390/ijerph18136932

Food Tank. 2023. "36 Organizations Helping Solve the Climate Crisis." https://foodtank.com/news/2020/10/36-organizations-helping-solve-the-climate-crisis/

Greenly. 2024. "What is the Average American Carbon Footprint and How to Reduce It?" https://greenly.earth/en-us/blog/company-guide/what-is-the-average-american-carbon-footprint-and-how-to-reduce-it

International Energy Agency. 2022. "The Future of Heat Pumps." https://www.iea.org/reports/the-future-of-heat-pumps/executive-summary

McKinsey. 2022. "Building decarbonization: How electric heat pumps could help reduce emissions today and going forward."

https://www.mckinsey.com/industries/electric-power-and-natural-gas/our-insights/building-decarbonization-how-electric-heat-pumps-could-help-reduce-emissions-today-and-going-forward

NREL (National Renewable Energy Laboratory). 2024. "Benefits of Heat Pumps Detailed in New NREL Report." https://www.nrel.gov/news/press/2024/benefits-of-heat-pumps-detailed-in-new-nrel-report.html

Perch Energy. 2024. "What Is the Average Carbon Footprint in the U.S.?" https://www.perchenergy.com/blog/environment/what-is-average-carbon-footprint-person-usa

Poore, J., and T. Nemecek. 2018. "Reducing food's environmental impacts through producers and consumers." Science 360(6392), 987-992. https://doi.org/10.1126/science.aaq0216

Rewiring America. 2024. "Heat pumps are appreciating climate assets." https://www.rewiringamerica.org/research/circuit-breakers/electrification-myths-heat-pump-efficiency

Ritchie, Hannah. 2023. "Interactive: What is the climate impact of eating meat and dairy?" Carbon Brief. https://interactive.carbonbrief.org/what-is-the-climate-impact-of-eating-meat-and-dairy/index.html

Robert Wood Johnson Foundation. 2024. "Five Lessons on Community-Driven Climate Action." https://www.rwjf.org/en/insights/blog/2024/01/five-lessons-on-community-driven-climate-action.html

Rocky Mountain Institute. 2023. "The Economics of Electrifying Buildings." https://rmi.org/insight/the-economics-of-electrifying-buildings/

Taiebat, Morteza, and Ming Xu. 2019. "5 charts show how your household drives up global greenhouse gas emissions." PBS News Weekend. https://www.pbs.org/newshour/science/5-charts-show-how-your-household-drives-up-global-greenhouse-gas-emissions

United Nations. 2022. "Small solutions, big impacts: 5 community-based projects tackling climate change." UN News. https://news.un.org/en/story/2022/04/1117122

United Nations. 2023. "Food and Climate Change: Healthy diets for a healthier planet." https://www.un.org/en/climatechange/science/climate-issues/food

U.S. Department of Energy. 2023. "Heat Pump Systems." https://www.energy.gov/energysaver/heat-pump-systems

U.S. Energy Information Administration. 2023. "U.S. Energy-Related Carbon Dioxide Emissions, 2023." https://www.eia.gov/environment/emissions/carbon/

Weber, C.L., and H.S. Matthews. 2008. "Food-miles and the relative climate impacts of food choices in the United States." Environmental Science & Technology 42(10): 3508-3513. https://doi.org/10.1021/es702969f

Chapter 12 Works Cited

Clayton, S., Bamberg, S., Reese, G., Fujii, S., & Schultz, P. W. (2023). Psychology and climate change: The diverse impacts on human wellbeing and the role of psychology in addressing them. Current Opinion in Psychology, 50, 101582. https://doi.org/10.1016/j.copsyc.2023.101582

Climate-XChange. (2020). Communicating the climate crisis. Retrieved from https://climate-xchange.org/communicating-the-climate-crisis/

Gifford, R. (2011). The dragons of inaction: Psychological barriers that limit climate change mitigation and adaptation. American Psychologist, 66(4), 290–302. https://doi.org/10.1037/a0023566

Hornsey, M. J. (2021). Why facts are not enough: Understanding and managing the motivated rejection of science. Current Directions in Psychological Science, 30(6), 535-542. https://doi.org/10.1177/09637214211037868

IPCC. (2022). Climate Change 2022: Impacts, Adaptation and Vulnerability. Contribution of Working Group II to the Sixth Assessment Report of the Intergovernmental Panel on Climate Change. Cambridge University Press, Cambridge, UK and New York, NY, USA. https://www.ipcc.ch/report/ar6/wg2/

Korteling, J. E., de Boer, L. C., Pak, R., Tinard, L., Borghuis, J., & Toet, A. (2023). Cognitive bias and how to improve sustainable decision making. Frontiers in Psychology, 14, 1129835. https://doi.org/10.3389/fpsyg.2023.1129835

Marshall, N., Adger, W. N., Benham, C., Brown, K., Curnock, M. I., Gurney, G. G., Marshall, P., Pert, P. L., & Thiault, L. (2021). Reef grief: investigating the relationship between place meanings and place change on the Great Barrier Reef, Australia. Sustainability Science, 16, 1261-1675. https://doi.org/10.1007/s11625-019-00666-z

NASA. (2024). Mitigation and adaptation. NASA Science. Retrieved from https://science.nasa.gov/climate-change/adaptation-mitigation/

NOAA. (2024). Climate. National Oceanic and Atmospheric Administration. Retrieved from https://www.noaa.gov/climate

Ojala, M. (2021). To worry or not to worry? The importance of hope, worry and anxiety for environmental engagement. Anxiety, Stress, & Coping, 34(3), 313-329. https://doi.org/10.1080/10615806.2020.1790453

Pihkala, P. (2022). Climate anxiety: What it is and what to do with it. Current Opinion in Environmental Sustainability, 57, 101209. https://doi.org/10.1016/j.cosust.2022.101209

Yale Program on Climate Change Communication. (2021). Talking climate with those holding different worldviews. Yale Climate Connections. Retrieved from https://yaleclimateconnections.org/2021/06/talking-climate-with-those-holding-different-worldviews/

Zhao, J., & Luo, Y. (2021). *A framework to address cognitive biases of climate change.* *Neuron, 109(22), 3548-3551. https://doi.org/10.1016/j.neuron.2021.08.029*

Chapter 13 Works Cited

Adger, W. N., Butler, C., & Walker-Springett, K. (2017). Moral reasoning in adaptation to climate change. Environmental Politics, 26(3), 371-390. https://doi.org/10.1080/09644016.2017.1287624

ASCE (American Society of Civil Engineers). (2021). Infrastructure Report Card: A Comprehensive Assessment of America's Infrastructure. https://infrastructurereportcard.org/

Busby, J. W., Baker, K., Bazilian, M. D., Gilbert, A. Q., Grubert, E., Rai, V., Rhodes, J. D., Shidore, S., Smith, C. A., & Webber, M. E. (2021). Cascading risks: Understanding the 2021 winter blackout in Texas. Energy Research & Social Science, 77, 102106. https://doi.org/10.1016/j.erss.2021.102106

Clayton, S., Manning, C. M., Krygsman, K., & Speiser, M. (2017). Mental Health and Our Changing Climate: Impacts, Implications, and Guidance. American Psychological Association, and ecoAmerica. https://www.apa.org/news/press/releases/2017/03/mental-health-climate.pdf

Ebi, K. L., Capon, A., Berry, P., Broderick, C., de Dear, R., Havenith, G., Honda, Y., Kovats, R. S., Ma, W., Malik, A., Morris, N. B., Nybo, L., Seneviratne, S. I., Vanos, J., & Jay, O. (2021). Hot weather and heat extremes: health risks. The Lancet, 398(10301), 698-708. https://doi.org/10.1016/S0140-6736(21)01208-3

EPA (Environmental Protection Agency). (2023). Climate Change Impacts on Water Resources. https://www.epa.gov/climate-indicators/water-resources

FAO (Food and Agriculture Organization). (2021). The State of Food and Agriculture 2021: Making agrifood systems more resilient to shocks and stresses. Rome. https://doi.org/10.4060/cb4476en

FAO (Food and Agriculture Organization). (2023). Regenerative agriculture: An opportunity for building resilience in the face of climate change? Rome. https://www.fao.org/documents/card/en/c/cc3771en

IPBES (Intergovernmental Science-Policy Platform on Biodiversity and Ecosystem Services). (2019). Global assessment report on biodiversity and ecosystem services of the Intergovernmental Science-Policy Platform on Biodiversity and Ecosystem Services. E. S. Brondizio, J. Settele, S. Díaz, and H. T. Ngo (editors). IPBES secretariat, Bonn, Germany. https://doi.org/10.5281/zenodo.3831673

IPCC (Intergovernmental Panel on Climate Change). (2022). Climate Change 2022: Impacts, Adaptation and Vulnerability. Contribution of Working Group II to the Sixth Assessment Report of the Intergovernmental Panel on Climate Change. Cambridge University Press. https://www.ipcc.ch/report/ar6/wg2/

Kharouba, H. M., Ehrlén, J., Gelman, A., Bolmgren, K., Allen, J. M., Travers, S. E., & Wolkovich, E. M. (2018). Global shifts in the phenological synchrony of species interactions over recent decades. Proceedings of the National Academy of Sciences, 115(20), 5211-5216. https://doi.org/10.1073/pnas.1714511115

Mora, C., Dousset, B., Caldwell, I. R., Powell, F. E., Geronimo, R. C., Bielecki, C. R., Counsell, C. W. W., Dietrich, B. S., Johnston, E. T., Louis, L. V., Lucas, M. P., McKenzie, M. M., Shea, A. G., Tseng, H., Giambelluca, T. W., Leon, L. R., Hawkins, E., & Trauernicht, C. (2017). Global risk of deadly heat. Nature Climate Change, 7(7), 501-506. https://doi.org/10.1038/nclimate3322

National Research Council. (2010). Adapting to the Impacts of Climate Change. The National Academies Press. https://doi.org/10.17226/12783

Oppenheimer, M., Glavovic, B. C., Hinkel, J., van de Wal, R., Magnan, A. K., Abd-Elgawad, A., Cai, R., Cifuentes-Jara, M., DeConto, R. M., Ghosh, T., Hay, J., Isla, F., Marzeion, B., Meyssignac, B., & Sebesvari, Z. (2019). Sea Level Rise and Implications for Low-Lying Islands, Coasts and Communities. In: IPCC Special Report on the Ocean and Cryosphere in a Changing Climate. Cambridge University Press. https://www.ipcc.ch/srocc/chapter/chapter-4-sea-level-rise-and-implications-for-low-lying-islands-coasts-and-communities/

Pelling, M., O'Brien, K., & Matyas, D. (2015). Adaptation and transformation. Climatic Change, 133(1), 113-127. https://doi.org/10.1007/s10584-014-1303-0

Siders, A. R., Hino, M., & Mach, K. J. (2019). The case for strategic and managed climate retreat. Science, 365(6455), 761-763. https://doi.org/10.1126/science.aax8346

USGCRP (U.S. Global Change Research Program). (2018). Impacts, Risks, and Adaptation in the United States: Fourth National Climate Assessment, Volume II. Reidmiller, D.R., C.W. Avery, D.R. Easterling, K.E. Kunkel, K.L.M. Lewis, T.K. Maycock, and B.C. Stewart (eds.). U.S. Global Change Research Program, Washington, DC, USA. https://nca2018.globalchange.gov/

WHO (World Health Organization). (2021). COP26 special report on climate change and health: the health argument for climate action. World Health Organization. https://www.who.int/publications/i/item/9789240036727

Chapter 14 Works Cited

Amnesty International. "Climate Activists Greta Thunberg and the Fridays for Future Movement Honoured with Top Amnesty International Award." Amnesty International, 2019, www.amnesty.org/en/latest/press-release/2019/06/greta-thunberg-and-fridays-for-future-win-ambassador-of-conscience-2019-award/.

Common Dreams. "Juliana v. US: Supreme Court Decision Brings 10-Year Climate Case to an End." Common Dreams, 2025, www.commondreams.org/news/juliana-vs-us.

Earth.Org. "Fridays for Future: How Young Climate Activists Are Making Their Voices Heard." Earth.Org, 2023, earth.org/fridays-for-future/.

Fisher, Dana R., and Sohana Nasrin. "Climate Activism and Its Effects." Wiley Interdisciplinary Reviews: Climate Change, vol. 12, no. 1, 2021, e683.

Fridays For Future. "Fridays For Future International Climate Movement." Fridays For Future, 2023, fridaysforfuture.org/.

Global Center on Adaptation. "Why We Need More Research-Based Youth Activism, Especially in Climate Adaptation." Global Center on Adaptation, 2024, gca.org/why-we-need-more-research-based-youth-activism-especially-in-climate-adaptation/.

Grist. "The World's Biggest Youth Climate Lawsuit Lost in Court, but It 'Changed the World'." Grist, 2025, grist.org/justice/juliana-v-united-states-climate-lawsuit-supreme-court-changed-the-world/.

MDPI. "Youth Mobilization to Stop Global Climate Change: Narratives and Impact." Sustainability, 2020, www.mdpi.com/2071-1050/12/10/4127.

Our Children's Trust. "Juliana v. United States." Our Children's Trust, 2025, www.ourchildrenstrust.org/juliana-v-us.

PubMed Central. "Children, Adolescents, and Youth Pioneering a Human Rights-Based Approach to Climate Change." PMC, 2021, pmc.ncbi.nlm.nih.gov/articles/PMC8694303/.

Rapid Transition Alliance. "The Sunrise Movement: How a US Grassroots Youth Movement Helped Set the National Climate Agenda for Rapid Change." Rapid Transition Alliance, 2021, rapidtransition.org/stories/the-sunrise-movement-how-a-us-grassroots-youth-movement-helped-set-the-national-climate-agenda-for-rapid-change/.

Resilience. "The Sunrise Movement: How a US Grassroots Youth Movement Helped Set the National Climate Agenda for Rapid Change." Resilience, 2021, www.resilience.org/stories/2021-03-11/the-sunrise-movement-how-a-us-grassroots-youth-movement-helped-set-the-national-climate-agenda-for-rapid-change/.

Sunrise Movement. "Green New Deal." Sunrise Movement, 2023, www.sunrisemovement.org/green-new-deal/.

Sustainability Science. "Perceived Impacts of the Fridays for Future Climate Movement on Environmental Concern and Behaviour in Switzerland." Sustainability Science, 2023, link.springer.com/article/10.1007/s11625-023-01348-7.

Taylor & Francis Online. "Youth Climate Activists Meet Environmental Governance: Ageist Depictions of the FFF Movement and Greta Thunberg in German Newspaper Coverage." Taylor & Francis Online, 2020, www.tandfonline.com/doi/full/10.1080/17447143.2020.1745211.

The Conversation. "Fridays for Future: How the Young Climate Movement Has Grown Since Greta Thunberg's Lone Protest." The Conversation, 2020, theconversation.com/fridays-for-future-how-the-young-climate-movement-has-grown-since-greta-thunbergs-lone-protest-144781.

Time. "The Sunrise Movement Rethinks Approach to Creating Change." Time, 2022, time.com/6158322/sunrise-movement-climate-activism-struggles/.

Zero Hour. "This Is Zero Hour | A Youth-led Movement." Zero Hour, 2024, thisiszerohour.org/.

Chapter 15 Works Cited

Anderson, A.A. 2021. "Experiencing Climate Change Through Everyday Life: Exploring Small Talk as a Climate Communication Tool." Journal of Science Communication 20(1): A07.

Bamberg, S., J. Rees, and M. Schulte. 2023. "Environmental Protection Through Societal Change: What We Know About Collective Climate Action and What We Need to Find Out." Current Opinion in Psychology 48: 101410.

Bollinger, B., and K. Gillingham. 2012. "Peer Effects in the Diffusion of Solar Photovoltaic Panels." Marketing Science 31(6): 900-912.

Bonvillian, W.B., and S. Sarma. 2021. "Accelerating Climate Change Solution Deployment." Issues in Science and Technology 37(3): 26-30.

Brand, C., M. Anable, M. Kränkel, and T. Preston. 2021. "The Climate Change Mitigation Effects of Daily Active Travel in Cities." Transportation Research Part D: Transport and Environment 93: 102764.

Breuer, C., S. Hüffmeier, and J. Hertel. 2021. "Does Working from Home Reduce CO2 Emissions? An Analysis of Travel Patterns as a Consequence of the COVID-19 Lockdown." Journal of Industrial Ecology 25(6): 1485-1498.

Burger, M., J. Wentz, and R. Horton. 2020. "The Law and Science of Climate Change Attribution." Columbia Journal of Environmental Law 45(1): 57-240.

Capstick, S., L. Whitmarsh, W. Poortinga, N. Pidgeon, and P. Upham. 2020. "Climate Change Citizenships and Participation: A Review and Research Agenda." Wiley Interdisciplinary Reviews: Climate Change 11(4): e588.

Carbon Trust. 2023. "Measure and Report Environmental Impact." Carbon Trust. https://www.carbontrust.com/what-we-do/measure-and-evaluate/measure-and-report-environmental-impact

CDP. 2023. "Supply Chain Program." CDP. https://www.cdp.net/en/supply-chain

Charter, M., and S. Keiller. 2014. "Grassroots Innovation and the Circular Economy: A Global Survey of Repair Cafés and Hackerspaces." The Centre for Sustainable Design, University for the Creative Arts.

Clayton, S., A. Carrico, and K. Steg. 2023. "The Psychology of Climate Change Communication: A Research-Based Guide for Climate Communicators." Journal of Environmental Psychology 83: 101872.

Cohen, B., A. Cowie, M. Babiker, et al. 2021. "Co-benefits of Mitigation." Chapter 12 in Climate Change 2022: Mitigation of Climate Change. Cambridge University Press.

Corner, A., H. Wang, and A. Roberts. 2020. "How to Talk About Climate Change: A Toolkit for Encouraging Collective Action." Climate Outreach.

DOE (Department of Energy). 2022. "Better Buildings Residential Network." Energy.gov. https://www.energy.gov/eere/better-buildings-residential-network/better-buildings-residential-network

DOE (Department of Energy). 2023. "Weatherization Assistance Program." Energy.gov. https://www.energy.gov/scep/wap/weatherization-assistance-program

Ebi, K.L., J. Hallegatte, T. Kram, et al. 2018. "Health, Wellbeing, and the Changing Structure of Communities." Chapter 5 in Global Warming of 1.5°C. Intergovernmental Panel on Climate Change.

Ellen MacArthur Foundation. 2021. "Universal Circular Economy Policy Goals." Ellen MacArthur Foundation.

EPA (Environmental Protection Agency). 2022. "Sustainable Materials Management in the Workplace." EPA.gov. https://www.epa.gov/smm/sustainable-materials-management-workplace

EPA (Environmental Protection Agency). 2023. "Home Energy Audits." Energy Star. https://www.energystar.gov/campaign/assessYourHome

FEMA (Federal Emergency Management Agency). 2023. "Ready: Community Preparedness." Ready.gov. https://www.ready.gov/community-preparedness

Fisher, D.R., and S. Nasrin. 2021. "Climate Activism and Its Effects." Wiley Interdisciplinary Reviews: Climate Change 12(1): e683.

Frenken, K., and J. Schor. 2017. "Putting the Sharing Economy into Perspective." Environmental Innovation and Societal Transitions 23: 3-10.

Geiger, N., J.T. Swim, and K. Glenna. 2021. "Spread the Green Word: A Social Community Perspective Into Environmentally Sustainable Behavior." Environment and Behavior 53(2): 138-170.

Gough, M.Z., and J. Accordino. 2013. "Public Gardens as Sustainable Community Development Partners: Motivations, Perceived Benefits, and Challenges." Urban Affairs Review 49(6): 851-887.

Heller, M.C. 2022. "Optimizing Climate Impact: A Guide for Individual Consumers." Annual Review of Environment and Resources 47: 1-25.

IPCC (Intergovernmental Panel on Climate Change). 2022. Climate Change 2022: Mitigation of Climate Change. Cambridge University Press.

IRENA (International Renewable Energy Agency). 2022. "Corporate Sourcing of Renewables: Market and Industry Trends." IRENA.

Kardan, O., P. Gozdyra, B. Misic, et al. 2015. "Neighborhood Greenspace and Health in a Large Urban Center." Scientific Reports 5: 11610.

Klöwer, M., D. Hopkins, M. Allen, and J. Higham. 2020. "An Analysis of Ways to Decarbonize Conference Travel After COVID-19." Nature 583: 356-359.

Kotcher, J., E. Maibach, W.T. Choi, et al. 2018. "Views of AAAS Scientists: Climate Change, Energy, Transportation, and Research Funding." Center for Climate Change Communication, George Mason University.

Markowitz, E.M., and A.F. Shariff. 2012. "Climate Change and Moral Judgement." Nature Climate Change 2: 243-247.

Martinez, D.F., and P. McManus. 2022. "Grassroots Climate Activism: Emerging Pathways of Climate Action." Environmental Politics 31(1): 126-149.

Nielsen. 2018. "Global Consumers Seek Companies That Care About Environmental Issues." Nielsen Insights, www.nielsen.com

NREL (National Renewable Energy Laboratory). 2022. "Community Solar." NREL.gov. https://www.nrel.gov/state-local-tribal/community-solar.html

Okvat, H.A., and A.J. Zautra. 2011. "Community Gardening: A Parsimonious Path to Individual, Community, and Environmental Resilience." American Journal of Community Psychology 47(3-4): 374-387.

Pielke Jr., R.A. 2018. "Opening Up the Climate Policy Envelope." Issues in Science and Technology 34(4): 30-36.

Plumer, B., and N. Popovich. 2019. "How Do You Plan for a City That Doesn't Exist Yet?" The New York Times, 11 June 2019.

Poore, J., and T. Nemecek. 2018. "Reducing Food's Environmental Impacts Through Producers and Consumers." Science 360(6392): 987-992.

Rainforest Action Network. 2023. "Banking on Climate Chaos: Fossil Fuel Finance Report 2023." Rainforest Action Network.

Ralph, K.M., C.T. Voulgaris, B.D. Taylor, et al. 2020. "Investigating the Influence of Employer Transportation Demand Management Strategies on Employee Travel Behavior." Transportation Research Part A: Policy and Practice 140: 132-141.

Rewiring America. 2023. "Electrify Everything in Your Home: A Guide to Comfy, Healthy, Carbon-Free Living." Rewiring America. https://www.rewiringamerica.org/electrify-home-guide

Robertson, J.L., and J. Barling. 2015. "The Role of Leadership in Promoting Workplace Pro-Environmental Behaviors." Oxford Handbook of Leadership and Organizations, Oxford University Press.

Rogers, T. 2020. "Ballot Initiatives: When Political Institutions Fail, Change Goes Directly to the People." Center for Public Impact.

Romero-Lankao, P., T. McPhearson, and D.J. Davidson. 2018. "The Food-Energy-Water Nexus and Urban Complexity." Nature Climate Change 8: 260-266.

Stuart, E., P. Larsen, J.C. Goldman, and D. Gilligan. 2020. "A Method to Estimate the Size and Remaining Market Potential of the U.S. ESCO (Energy Service Company) Industry." Energy Policy 77: 191-200.

UNESCO. 2022. "Learn for our Planet: Climate Change Education Around the World." UNESCO.

van Staalduinen, B.J., F. Khan, L. Gadag, and D. Renaud. 2022. "Legislative Staff Are Influenced by Constituent Contact: Evidence from a Field Experiment." Journal of Political Institutions and Political Economy 3(1): 1-26.

Vandenbergh, M.P., and J.M. Gilligan. 2017. Beyond Politics: The Private Governance Response to Climate Change. Cambridge University Press.

Wagner, G., T. Kåberger, S. Olai, et al. 2020. "Energy Policy: Push Renewables to Spur Carbon Pricing." Nature 525: 27-29.

Wang, D. 2018. "Local Environmental Governance and Public Participation." Chinese Political Science Review 3: 28-41.

Willett, W., J. Rockström, B. Loken, et al. 2019. "Food in the Anthropocene: The EAT–Lancet Commission on Healthy Diets from Sustainable Food Systems." The Lancet 393(10170): 447-492.

Wynes, S., and K.A. Nicholas. 2017. "The Climate Mitigation Gap: Education and Government Recommendations Miss the Most Effective Individual Actions." Environmental Research Letters 12(7): 074024.

Appendix B: Comprehensive Glossary: The Climate Dictionary

A

Adaptation: Adjusting to actual or expected climate effects to moderate harm or exploit beneficial opportunities. Like upgrading your umbrella collection when you realize it's going to rain forever, not just occasionally. Can be reactive (after impacts) or anticipatory (before impacts).

Aerosols: Tiny particles suspended in the atmosphere that can reflect sunlight back to space or absorb it. Some cool the planet (sulfates from volcanoes) while others warm it (black carbon from combustion). Nature's mood lighting system for Earth.

Afforestation: Planting new forests where there were none before. The climate equivalent of getting hair plugs for a previously bald Earth.

Agroforestry: Agricultural system combining trees with crops or livestock. Nature's version of multitasking—trees provide shade while fixing nitrogen while preventing erosion while producing fruit while making everything look prettier.

Albedo: The measure of how much solar radiation is reflected by a surface. Ice has high albedo (reflects lots of sunlight), while oceans have low albedo (absorb lots of sunlight). The planetary equivalent of wearing white versus black on a summer day.

Anthropocene: Proposed geological epoch dating from when human activities started to significantly impact Earth's geology and ecosystems. Basically when humans got promoted from "just another species" to "geological force of nature"—unfortunately not in a cool superhero way.

Anthropogenic: Resulting from human activities. When it's definitely us, not volcanoes or sunspots or "natural cycles," causing the problem.

Arctic Amplification: The phenomenon where the Arctic warms faster than the global average. Like when your feet get disproportionately cold compared to the rest of your body, except in reverse, and for the planet.

Atmospheric Lifetime: The average time a gas remains in the atmosphere. CO_2 has a lifetime of centuries, while methane sticks around for about 12 years. Some greenhouse gases are like dinner guests who know when to leave; others are like in-laws who move into your basement.

Atmospheric River: Long, narrow region in the atmosphere that transports water vapor outside of the tropics. Like a fire hose of moisture pointed at your previously drought-stricken region, delivering six months of rain in three days.

Attribution Science: Field of study determining the extent to which climate change influenced a specific extreme weather event. The scientific version of figuring out whether your teenager or their friend broke the lamp, except with supercomputers and peer review.

B

Baseline: Reference period against which changes are measured. In climate science, often 1850-1900 pre-industrial levels. Like the "before" picture in a weight loss ad, but for planetary temperature.

Biodiversity: The variety of life in a particular habitat or ecosystem. Mother Nature's insurance policy—the more species we have, the better our chances that some will adapt to our climate mess.

Biochar: Charcoal produced from plant matter and stored in soil to remove carbon dioxide from the atmosphere. Like giving the soil a charcoal face mask that lasts for centuries.

Biofuel: Fuel derived directly from living matter. Like converting your leftover french fry oil into something that makes your car smell like a fast food restaurant.

Biomass: Organic material from living or recently living organisms used as fuel. Technically that log in your fireplace is solar power—just stored for a few decades in wood form.

Blue Carbon: Carbon captured by the world's ocean and coastal ecosystems. The ocean's way of saying, "I've been offsetting your emissions this whole time and you didn't even send a thank-you note."

Business As Usual (BAU): Scenario for future patterns of activity that assumes that there will be no significant change in people's attitudes and priorities. The climate equivalent of ignoring that "check engine" light on your dashboard.

C

Cap and Trade: System where governments set a limit (cap) on certain emissions and companies can buy or sell (trade) permits to emit these gases. Like musical chairs for pollution rights, except when the music stops, companies can still buy chairs from each other.

Carbon Budget: The cumulative amount of carbon dioxide emissions permitted to maintain a chance of limiting warming to a specific level. Like a financial budget except when you overspend, coastal cities flood and ecosystems collapse.

Carbon Capture and Storage (CCS): Technology that captures CO_2 emissions and stores them underground rather than releasing them into the atmosphere. Like sweeping climate change under the geological rug, but in a good way.

Carbon Cycle: The process by which carbon travels from the atmosphere to the Earth and back to the atmosphere. Nature's own recycling program for carbon, which was working fine until we dug up fossil fuels and broke the return policy.

Carbon Dioxide (CO_2): A greenhouse gas produced by burning fossil fuels, forest fires, and respiration. The Leonardo DiCaprio of greenhouse gases—gets all the publicity while methane quietly does more damage per molecule.

Carbon Dioxide Equivalent (CO_2e): Metric measure used to compare emissions from various greenhouse gases based on their global-warming potential. The "let's convert everything to one currency" approach to greenhouse gases.

Carbon Footprint: The total greenhouse gas emissions caused directly and indirectly by an individual, organization, or product. The climate guilt score you try to reduce while still occasionally ordering same-day delivery.

Carbon Neutral: Achieving net-zero carbon dioxide emissions by balancing emissions with carbon removal or simply eliminating carbon emissions altogether. Like maintaining your weight while still eating cake—theoretically possible, but requires a lot of offsetting effort.

Carbon Offset: Reduction in emissions of carbon dioxide or other greenhouse gases made to compensate for emissions made elsewhere. The climate equivalent of doing extra credit to make up for bombing that midterm.

Carbon Pricing: Making emitters pay for their carbon pollution, either through a carbon tax or cap-and-trade system. Putting a price tag on something we've been getting for free, like suddenly being charged for those "complimentary" hotel breakfasts.

Carbon Sequestration: The process of capturing and storing atmospheric carbon dioxide. Basically giving carbon a nice, long timeout underground or in plants rather than letting it roam freely in the atmosphere.

Carbon Sink: Anything that absorbs more carbon than it releases, like forests, soil, and oceans. Nature's carbon vacuum cleaners, quietly doing their job until we cut them down, plow them up, or warm them too much.

Circular Economy: Economic system aimed at eliminating waste by continually reusing resources. What your grandparents practiced without calling it anything fancy—"Why throw it away when you could fix it, repurpose it, or give it to your cousin?"

Climate: The average weather conditions in a place over a long period (typically 30+ years). Not what's happening outside your window right now, Karen, so put away that snowball you brought to Congress.

Climate Justice: Framework addressing the ethical dimensions of climate change, considering historical responsibilities and the unequal impacts on vulnerable populations. Because some people drove the gas-guzzling SUV but others are getting the speeding ticket.

Climate Model: Computer simulation of the climate system used to study past changes and project future climate. Incredibly complex mathematical representations that still get criticized when they're off by half a degree after predicting 30 years into the future.

Climate Sensitivity: The amount of global surface warming that will occur after a doubling of atmospheric CO_2. Climate science's version of "how badly will this hurt?", measured in degrees Celsius.

Community Solar: Solar power installations where multiple participants share costs and benefits. Like a community garden, but instead of zucchini, you harvest clean electricity and lower utility bills.

Coral Bleaching: When corals expel their symbiotic algae due to stress, particularly from high water temperatures, turning white and often dying. Like if you got so stressed you ejected your digestive system—not a sustainable survival strategy.

D

Decarbonization: Reducing carbon dioxide emissions from an energy system. The process of putting our carbon addiction through rehab.

Deforestation: Clearing of forests for agriculture, urban development, or other purposes. Like giving the Earth an unwanted haircut that hurts its ability to breathe.

Dendrochronology: Scientific method of dating based on the analysis of tree ring patterns. Nature's own barcode system, recording the climate story one growth ring at a time.

Desertification: The degradation of land in dry areas due to climate change and human activities. When formerly productive land starts looking like a Mars rover photo.

Divestment: The removal of investment capital from stocks, bonds, or funds tied to fossil fuel companies. Telling fossil fuels, "It's not me, it's you," and taking your money elsewhere.

Drawdown: The point at which greenhouse gas levels in the atmosphere begin to decline. The climate equivalent of reaching the top of the roller coaster and finally starting the descent.

Drought: Period of abnormally dry weather sufficiently prolonged to cause serious hydrological imbalance. When your lawn turns crunchy, your reservoir becomes a mud pit, and your water bill makes you consider showering with bottled water.

E

Earth System Science: Interdisciplinary study of Earth as a system of interacting physical, chemical, and biological processes. The "it's all connected" approach to planetary science that makes climate scientists sound like hippies even when they're using supercomputers.

Eco-anxiety: Chronic fear of environmental doom. When checking the climate news feels like doomscrolling but with more scientific graphs.

El Niño/La Niña: Periodic warming (El Niño) and cooling (La Niña) of the central and eastern Pacific Ocean that affects weather patterns worldwide. Nature's climate DJs, mixing up weather patterns and confusing meteorologists since observations began.

Electrification: The process of replacing technologies that use fossil fuels with those that use electricity. Trading your gas-guzzling appliances for a houseful of devices that all need to be charged, but don't poison the air.

Energy Efficiency: Using less energy to provide the same service. The art of getting more bang for your energy buck without sitting in the dark or freezing in winter.

Energy Return on Investment (EROI): Ratio of energy delivered by a process to energy used directly and indirectly in that process. The "is this even worth it?" calculation for energy sources.

Environmental Justice: Fair treatment and meaningful involvement of all people with respect to environmental regulations and policies. Because pollution shouldn't target neighborhoods based on income or race, even though it historically has.

Evapotranspiration: Combined process of water surface evaporation and plant transpiration. Plants sweating and water evaporating, all in one fancy scientific term that's awkwardly long for Scrabble.

F

Feedback Loop: A process where the effects of climate change either amplify (positive feedback) or diminish (negative feedback) the rate of change. When climate change causes effects that cause more climate change —like a bad relationship where fighting about fighting becomes its own problem.

Food Security: When all people, at all times, have physical, social, and economic access to sufficient, safe, and nutritious food. What becomes threatened when you can no longer grow crops where you used to because your farmland is underwater, on fire, or baking in an endless drought.

Fossil Fuels: Natural fuels formed in the geological past from the remains of living organisms. Basically, we're burning the decomposed remains of plants and dinosaurs to power our Netflix binges.

Fracking: Hydraulic fracturing—process of injecting liquid at high pressure into rocks to extract oil or gas. Like giving the Earth an unwanted pressure-washing from the inside.

G

Geoengineering: Large-scale technological interventions in Earth's climate system intended to counteract global warming. Plan B that sounds like science fiction until Plan A (emissions reduction) keeps failing. Includes ideas like: "What if we just block some sunlight?" and "Can we make the ocean absorb more carbon?"

Glacier: Persistent body of dense ice constantly moving under its own weight. Earth's freshwater savings accounts that we're breaking into at an alarming rate.

Global Warming: Long-term heating of Earth's climate system due to human activities, primarily fossil fuel burning. The original climate term before it was rebranded to "climate change" because some people heard "warming" and thought, "That sounds nice!"

Global Warming Potential (GWP): Measure of how much energy emissions of one ton of a gas will absorb over time, relative to emissions of one ton of CO_2. The "how bad is this compared to CO_2?" scale for greenhouse gases.

Green Building: Building designed to reduce environmental impact through efficient use of resources. When your building is environmentally friendly, not just painted green.

Green New Deal: Proposed package combining climate action with economic stimulus and social justice initiatives. Like FDR's New Deal, but with solar panels and fewer dams.

Greenhouse Effect: Process by which radiation from a planet's atmosphere warms the planet's surface. Naturally keeps Earth habitable, but we've cranked it up to "uncomfortably warm" levels.

Greenhouse Gas (GHG): Gas that absorbs and emits radiant energy within the thermal infrared range, causing the greenhouse effect. Including CO_2, methane, nitrous oxide, and water vapor. The atmospheric blanket that's getting uncomfortably thick.

Greenwashing: Misleading claims about the environmental benefits of a product, service, or company policy. When a company puts a picture of a forest on their plastic packaging and calls it "eco-friendly."

Grid: Interconnected network delivering electricity from producers to consumers. The biggest and most complex machine humans have ever built, which we now need to completely redesign for renewable energy.

H

Heat Dome: Persistent high-pressure system that traps hot air over a region for days or weeks. Like putting a lid on a pot, except the pot is your city and the lid is trapping heat instead of steam.

Heat Pump: Device that transfers heat from a cooler space to a warmer space using mechanical energy. The HVAC system that will save the world if we can just install enough of them.

Heat Wave: Period of abnormally and uncomfortably hot weather. When it's so hot that everyone becomes collectively obsessed with temperature and can't talk about anything else.

Hockey Stick Graph: Chart showing relatively stable global temperatures over the past 1,000 years followed by sharp warming in the 20th century. Resembles a hockey stick lying flat with the blade pointing upward. The graph climate deniers hate more than actual hockey players hate the opposing team in the Stanley Cup finals.

Hydrological Cycle: The continuous movement of water on, above, and below Earth's surface. Nature's ultimate recycling program for water, which climate change is currently reprogramming.

I

Ice Core: Cylinder of ice drilled from an ice sheet or glacier, used to study past climate conditions. Earth's time capsules, with atmospheric records going back hundreds of thousands of years. Like tree rings but colder and spanning much longer timescales.

Ice Sheet: Mass of glacial ice covering surrounding terrain and greater than 50,000 square kilometers. Earth's air conditioners and water storage systems that are currently melting like ice cream on a hot sidewalk.

Intergovernmental Panel on Climate Change (IPCC): UN body evaluating the science related to climate change. Thousands of scientists volunteering their time to write reports that policymakers often praise and then ignore.

Invasive Species: Non-native organism that causes ecological or economic harm in a new environment. The unwanted tourists of the biological world who decide to stay permanently and rearrange the furniture.

J

Jet Stream: Fast-flowing, narrow air current in the atmosphere that influences weather patterns. The atmospheric highway that's becoming increasingly erratic due to climate change, like a roller coaster designed by someone having a bad day.

Just Transition: Transitioning to a low-carbon economy in a way that's fair to everyone, particularly those whose livelihoods depend on high-carbon industries. Making sure that when we switch from fossil fuels, coal miners don't just get a pamphlet saying "Learn to code."

K

Kilowatt-hour (kWh): Unit of energy equivalent to one kilowatt of power sustained for one hour. What your utility charges you for, and what renewable energy is making increasingly cheaper.

Kyoto Protocol: International treaty extending the 1992 UN Framework Convention on Climate Change, committing countries to reduce greenhouse gas emissions. The climate agreement everyone remembers fondly now that newer agreements aren't doing much better.

L

La Niña: Period of cooler-than-normal sea surface temperatures in the central and eastern tropical Pacific Ocean that impacts global weather patterns. El Niño's chillier sister who brings different but equally disruptive weather patterns.

Land Use Change: Human modification of land for various purposes, often converting natural landscapes to human uses. When forests become farms, farms become suburbs, and everyone wonders why wildlife keeps showing up in their backyard.

Levelized Cost of Energy (LCOE): Measure of the average net present cost of electricity generation over a plant's lifetime. The "real cost" calculation that increasingly shows renewables beating fossil fuels at their own game.

Little Ice Age: Period of cooling that occurred after the Medieval Warm Period (approximately 1300-1850 CE). When Europeans could ice skate on the Thames and hold frost fairs on frozen rivers—activities their descendants won't be experiencing again anytime soon.

M

Managed Retreat: Planned relocation of people and assets away from areas vulnerable to climate impacts. The climate version of "know when to fold 'em" when keeping communities in the same location becomes untenable.

Maunder Minimum: Period of unusually low solar activity between approximately 1645 and 1715. What climate deniers wish was happening now instead of human-caused warming.

Medieval Warm Period: Warm period in the North Atlantic region lasting from c. 950 to c. 1250. Regional warming that climate skeptics like to mention while conveniently forgetting it wasn't global like current warming.

Megadrought: Severe drought lasting multiple decades. When you've been in a drought so long that being in a drought becomes your regional identity.

Methane (CH_4): Potent greenhouse gas with approximately 84-86 times the warming potential of CO_2 over a 20-year period. The greenhouse gas equivalent of that friend who doesn't drink often but goes wild at parties.

Microgrid: Localized energy system that can operate independently from the traditional grid. The energy version of being able to make your own coffee when the local café closes.

Mitigation: Efforts to reduce or prevent greenhouse gas emissions. Addressing the cause rather than the symptoms of climate change. Like turning off the faucet instead of just mopping up the overflowing sink.

N

Natural Gas: Primarily methane, often described as a "bridge fuel" between coal and renewables. The fossil fuel that got better PR than the others but is still a major source of methane emissions.

Net-Zero: Achieving a balance between greenhouse gas emissions produced and removed from the atmosphere. The climate equivalent of maintaining your weight while still occasionally enjoying ice cream.

NIMBY: Acronym for "Not In My Back Yard," describing opposition to local developments perceived as undesirable. What happens when everyone wants renewable energy but nobody wants to see wind turbines from their porch.

Nitrous Oxide (N_2O): Greenhouse gas commonly emitted from agricultural activities and fossil fuel combustion. The "laughing gas" that's nothing to laugh about when it comes to warming potential.

O

Ocean Acidification: Decrease in the pH of Earth's oceans caused by uptake of atmospheric CO_2. The "evil twin" of climate change that's making life particularly difficult for anything with a shell.

Ocean Current: Continuous, directed movement of seawater generated by forces like wind, temperature, and salinity differences. Earth's conveyor belts that distribute heat around the planet and are currently being reprogrammed by climate change.

Ozone Layer: Region of Earth's stratosphere that absorbs most of the Sun's ultraviolet radiation. The planetary sunscreen that we almost destroyed with CFCs and then actually managed to save through international cooperation.

P

Paris Agreement: International treaty on climate change adopted in 2015, aiming to limit global warming to well below 2°C above pre-industrial levels. The climate pact that everyone celebrated signing and then largely failed to implement.

Particulate Matter: Mixture of solid particles and liquid droplets in the air, often from burning fossil fuels. What makes your lungs and white laundry turn black in heavily polluted cities.

Passive House: Ultra-energy-efficient building standard requiring minimal energy for heating and cooling. Buildings so well-insulated they can nearly heat themselves with body heat and appliances. The introverts of the architectural world.

Permafrost: Ground that remains completely frozen for at least two consecutive years. Nature's freezer for carbon and methane that climate change is unplugging.

Photovoltaic (PV): Conversion of light into electricity using semiconducting materials. Solar panels that turn sunshine into Netflix power.

Positive Feedback Loop: Process where the effects of climate change amplify the rate of change. Like when melting ice reveals darker water that absorbs more heat that melts more ice. The "this is fine" dog in the burning room meme, but for planetary systems.

Pre-industrial Levels: Usually refers to greenhouse gas concentrations or temperatures before widespread industrialization, typically before 1750. The "before" picture in Earth's climate makeover story.

R

Radiative Forcing: Change in energy flux in the atmosphere caused by natural or human-induced factors. The climate's thermostat that we've been turning up through greenhouse gas emissions.

REDD+: "Reducing Emissions from Deforestation and forest Degradation" plus conservation and sustainable management. Paying countries to keep their forests standing instead of cutting them down.

Regenerative Agriculture: Farming practices that reverse climate change by rebuilding soil organic matter and restoring degraded soil biodiversity. Farming like your great-grandparents did, but with better science and fewer horse-drawn plows.

Renewable Energy: Energy collected from resources that are naturally replenished on a human timescale, such as sunlight, wind, rain, and geothermal heat. Power sources that don't need millions of years and dinosaur remains to replenish.

Resilience: Capacity of social, economic, and environmental systems to cope with hazardous events while maintaining essential function. The ability to take a punch from climate change and get back up.

S

Sea Level Rise: Increase in the level of the world's oceans due to global warming. The slow-motion disaster giving beachfront property a whole new meaning—underwater.

Sixth Mass Extinction: Ongoing extinction event driven by human activities. When we're losing species faster than nature can create new ones, and it's definitely our fault this time, not an asteroid's.

Solar Radiation Management: Type of geoengineering that would reflect sunlight back to space. The "maybe we should just dim the sun a bit" approach to climate change that sounds both brilliant and terrifying.

Storm Surge: Abnormal rise in seawater level during a storm, over and above the predicted tide. When a hurricane brings the ocean into your living room without an invitation.

Sustainable Development: Development that meets present needs without compromising future generations' ability to meet their own needs. The radical notion that our grandchildren might also want clean air, drinkable water, and a stable climate.

T

Tipping Point: Threshold that, when crossed, leads to large and often irreversible changes in a system. The climate equivalent of the straw that broke the camel's back, except the camel is a major Earth system like the Amazon rainforest or West Antarctic Ice Sheet.

Transition Risk: Financial risks related to the transition to a lower-carbon economy. When your coal company shares become worth less than the paper they're printed on because the world decided to stop burning rocks for energy.

Troposphere: Lowest region of Earth's atmosphere, where most weather occurs. The atmospheric layer where we're conducting our unintentional climate experiment.

U

Urban Heat Island: Urban area significantly warmer than its surrounding rural areas due to human activities and infrastructure. When cities trap heat like a black car seat in summer.

UV Index: Measurement of the strength of sunburn-producing ultraviolet radiation. The "how quickly will you turn into a lobster" scale that's becoming more relevant as ozone depletion and climate change progress.

V

Vector-borne Disease: Illness caused by pathogens transmitted by vectors like mosquitoes, ticks, and fleas. Diseases now appearing in new regions as climate change makes previously inhospitable areas suddenly welcoming to disease carriers.

Voluntary Carbon Market: Trading of carbon credits that allows companies or individuals to offset their emissions. The "pay someone else to plant trees while you keep driving your SUV" approach to climate action.

W

Water Cycle: Continuous movement of water within Earth and atmosphere through evaporation, condensation, precipitation, and runoff. Nature's water recycling system that climate change is putting on a more extreme setting.

Weather: State of the atmosphere at a particular place and time. What's happening outside your window right now, which is not the same as climate, Karen.

Weatherization: Making buildings more resistant to weather elements and more energy-efficient. The home improvement project that saves money, increases comfort, and fights climate change simultaneously.

Wet-bulb Temperature: Temperature read by a thermometer covered in water-soaked cloth over which air is passed. The "will humans literally cook in these conditions" measurement that's increasingly concerning climate scientists.

Z

Zero-carbon: Activities that produce no carbon emissions. The gold standard of climate-friendly behavior that's easier said than done.

Zero Waste: Conservation of resources through responsible production, consumption, reuse, and recovery without incineration or discharges that threaten the environment. When your trash can gets jealous of your compost bin and recycling center because they get all the action.

Zooplankton: Tiny animals that drift in oceans, seas, and bodies of fresh water. The ocean's tiny creatures that support marine food webs and are increasingly stressed by warming and acidification.

Appendix C: Book Index with Chapter References

Food security

- climate change threats to, Ch. 13
- adaptation strategies, Ch. 13
- global impacts on, Ch. 13

Fossil fuels

- carbon lock-in, Ch. 7
- divestment from, Ch. 9
- historical development, Ch. 3, Ch. 7
- subsidies for, Ch. 7
- transition away from, Ch. 7

G

Geoengineering

- carbon dioxide removal, Ch. 7
- ethical concerns, Ch. 7
- solar radiation management, Ch. 7

Glaciers

- as climate indicators, Ch. 4
- mass balance trends, Ch. 4
- retreat patterns, Ch. 4

Global warming

- causes of, Ch. 3
- definition of, Ch. 1
- historical trends, Ch. 1
- observed impacts, Ch. 1
- projections, Ch. 1

Greenhouse effect

- anthropogenic enhancement of, Ch. 1
- discovery and science of, Ch. 1
- natural process, Ch. 1

Greenhouse gases

- carbon dioxide, Ch. 1, Ch. 3
- comparison of warming potential, Ch. 3
- concentrations over time, Ch. 3
- methane, Ch. 1, Ch. 3
- natural vs. anthropogenic sources, Ch. 3
- nitrous oxide, Ch. 3

Grid modernization

- challenges and barriers, Ch. 10
- distributed energy integration, Ch. 10
- renewable energy integration, Ch. 10
- smart grid technologies, Ch. 10
- storage solutions, Ch. 10

H

Heat waves

- attribution to climate change, Ch. 6
- definition and trends, Ch. 6
- health impacts of, Ch. 13
- urban heat islands and, Ch. 6

Hockey stick graph

- controversy over, Ch. 8
- significance and evidence, Ch. 8
- validation of, Ch. 8

Home energy

- efficiency improvements, Ch. 11, Ch. 15
- electrification options, Ch. 15
- renewable integration, Ch. 11, Ch. 15

Methane (CH₄)

- agricultural sources, Ch. 1, Ch. 3
- comparison to CO_2, Ch. 3
- permafrost emissions, Ch. 1, Ch. 4

Microgrids

- advantages and applications, Ch. 10
- community resilience through, Ch. 10
- integration with main grid, Ch. 10

Mitigation

- carbon pricing strategies, Ch. 9
- definition and approaches, Ch. 7
- economy-wide strategies, Ch. 7
- relationship with adaptation, Ch. 13
- sectoral pathways, Ch. 7
- technological options, Ch. 7

N

Natural gas

- as transition fuel, Ch. 7
- environmental impacts of, Ch. 7
- methane leakage from, Ch. 3, Ch. 7

Nature-based solutions

- coastal protection, Ch. 13
- ecosystem services of, Ch. 13
- for carbon sequestration, Ch. 7, Ch. 13
- for flood control, Ch. 13

Net-zero emissions

- corporate commitments to, Ch. 9
- definition and pathways, Ch. 7
- timeline requirements, Ch. 7

O

Ocean acidification

- carbon chemistry of, Ch. 4
- ecosystem impacts, Ch. 4
- future projections, Ch. 4
- shellfish and coral effects, Ch. 4

Ocean circulation

- Atlantic Meridional Overturning Circulation, Ch. 4
- changes due to climate change, Ch. 4
- heat transport by, Ch. 4
- regional climate effects, Ch. 4

Ocean warming

- causes and patterns, Ch. 4
- coral bleaching from, Ch. 4
- ecosystem impacts, Ch. 4
- measurement of, Ch. 4

P

Paleoclimate

- ice core records, Ch. 8
- methods and proxies, Ch. 8
- sediment records, Ch. 8
- tree ring data, Ch. 8

Paris Agreement

- comparison to previous agreements, Ch. 7
- goals and mechanisms, Ch. 7
- implementation challenges, Ch. 7
- national determined contributions under, Ch. 7

Y

Youth climate activism

- accomplishments of, Ch. 14
- distinctive approaches of, Ch. 14
- future directions for, Ch. 14
- historical context of, Ch. 14
- intergenerational collaboration, Ch. 14
- lessons from, Ch. 14